Regulation of microRNAs

ADVANCES IN EXPERIMENTAL MEDICINE AND BIOLOGY

Recent Volumes in this Series

Volume 692
NEUROPEPTIDE SYSTEMS AS TARGETS FOR PARASITE AND PEST CONTROL
Edited by Timothy G. Geary and Aaron G. Maule

Volume 693
POST-TRANSCRIPTIONAL REGULATION BY STAR PROTEINS
Edited by Talila Volk and Karen Artzt

Volume 694
PROTEIN METABOLISM AND HOMEOSTASIS IN AGING
Edited by Nektarios Tavernarakis

Volume 695
THE CELL BIOLOGY OF STEM CELLS
Edited by Eran Meshorer and Kathrin Plath

Volume 696
SOFTWARE TOOLS AND ALGORITHMS FOR BIOLOGICAL SYSTEMS
Edited by Hamid R. Arabnia and Quoc-Nam Tran

Volume 697
HOT TOPICS IN INFECTION AND IMMUNITY IN CHILDREN VII
Edited by Andrew Pollard, Adam Finn, and Nigel Curtis

Volume 698
BIO-FARMS FOR NUTRACEUTICALS: FUNCTIONAL FOOD AND SAFETY CONTROL
BY BIOSENSORS
Maria Teresa Giardi, Giuseppina Rea and Bruno Berra

Volume 699
MCR 2009: PROCEEDINGS OF THE 4TH INTERNATIONAL CONFERENCE ON MULTI-
COMPONENT REACTIONS AND RELATED CHEMISTRY, EKATERINBURG, RUSSIA
Maxim A. Mironov

Volume 700
REGULATION OF MICRORNAS
Helge Großhans

Regulation of microRNAs

Edited by

Helge Großhans
Friedrich Miescher Institute for Biomedical Research, Basel, Switzerland

Springer Science+Business Media, LLC

Landes Bioscience

Springer Science+Business Media, LLC
Landes Bioscience

Printed in the USA.

Springer Science+Business Media, LLC, 233 Spring Street, New York, New York 10013, USA
http://www.springer.com

Please address all inquiries to the publishers:
Landes Bioscience, 1806 Rio Grande, Austin, Texas 78701, USA
Phone: 512/ 637 6050; FAX: 512/ 637 6079
http://www.landesbioscience.com

The chapters in this book are available in the Madame Curie Bioscience Database.
http://www.landesbioscience.com/curie

Regulation of microRNAs, edited by Helge Großhans. Landes Bioscience / Springer Science+Business Media, LLC dual imprint / Springer series: Advances in Experimental Medicine and Biology.

ISBN: 978-1-4419-7822-6

Library of Congress Cataloging-in-Publication Data

Regulation of microRNAs / edited by Helge Grosshans.
    p. ; cm. -- (Advances in experimental medicine and biology ; v. 700)
Includes bibliographical references and index.
ISBN 978-1-4419-7822-6
1. Small interfering RNA. 2. Genetic regulation. I. Grosshans, Helge. II. Series: Advances in experimental medicine and biology ; v. 700. 0065-2598
[DNLM: 1. MicroRNAs--genetics. 2. Gene Expression Regulation--genetics. W1 AD559 v.700 2010 / WY 105]
QP623.5.S63R445 2010
572.8'8--dc22

2010038832

PREFACE

When in 2001 the first draft versions of the human genome revealed that there were no more than 25,000 human genes, much soul searching resulted. How could a complicated human being develop and function with a gene set not much bigger than that of a worm, *Caenorhabditis elegans*?

Although the meaning of 'gene counts' is debatable when a single gene can give rise to a multitude of different gene products (and when in fact much of the 'inter-genic' genome appears to be transcribed), the apparent conundrum highlighted the importance of gene regulation in making complex organisms. It thus appears particularly appropriate that it was also in 2001 that microRNAs (miRNAs) were finding their way into the limelight. These regulatory RNAs, named for their small size of some 22 nucleotides, had been discovered in 1993 in *C. elegans*, but were initially considered a worm oddity and largely ignored. It was only when small RNA cloning efforts started to reveal hundreds of different miRNAs in a typical animal or plant genome that they were widely noticed. Today, it appears that hardly any cellular or developmental pathway has escaped the control that miRNAs exert by silencing target mRNAs through an antisense mechanism. Accordingly, miRNAs dysregulation contributes to numerous diseases, most notably diverse cancers.

Given this pervasiveness and importance of miRNA-mediated gene regulation, it should come as little surprise that miRNAs themselves are also highly regulated. However, the recent explosion of knowledge on this topic has been remarkable, providing a primary motivation for publication of this book. As miRNAs are transcribed by RNA polymerase II, the enzyme that also generates mRNAs, it was perhaps not unexpected that miRNA transcription would be subject to regulation, and we have willfully omitted this aspect from this monograph. However, what has been unexpected is the extent of post-transcriptional regulation of miRNAs that is illustrated in this book.

In the first chapter, René Ketting provides the background against which all of the regulatory processes occur by revealing the complex biogenesis and function of miRNAs and the related siRNAs. Akiko Hata and Brandi Davis then describe how SMAD proteins, generally known for their function in controlling transcription, reveal another side in regulating the processing of certain primary miRNA (pri-miRNA) transcripts by the RNase Drosha. Drosha-mediated processing of pri-miRNAs into the short precursor miRNAs (pre-miRNAs) is also modulated by the RNA-binding proteins hnRNP A1, as discussed by Javier Caceres and colleagues and KSRP (Michele Trabucchi et al). Whereas SMADs, hnRNP A1 and KSRP promote processing of specific pri-miRNAs, estrogen receptor alpha represses this biogenesis step as related by Shigeaki Kato and colleagues. Robinson Triboulet and Richard Gregory further reveal that the pri-miRNA processing complex undergoes autoregulation.

KSRP not only promotes processing of pri-miRNAs, but also the subsequent cleavage of pre-miRNAs by the RNase Dicer. Conversely, Lin28 was found to repress pri-miRNA as well pre-miRNA processing as discussed by Nicolas Lehrbach and Eric Miska. This inhibition involves 3' end uridylation of the pre-miRNA. Another RNA modification that occurs on miRNAs is adenosine-to-inosine editing, and Mary O'Connell and colleagues critically evaluate its incidence and how editing affects processing and functionality of miRNAs.

Gregory Wulczyn and colleagues discuss the Trim-NHL protein family whose members utilize diverse mechanisms to regulate miRNA levels and activity both positively and negatively. Nicole Meisner and Witold Filipowicz review HuR, an RNA-binding protein that regulates mRNAs through a number of mechanisms, including at least one instance in which HuR reverses miRNA-mediated mRNA silencing.

Finally, although mature miRNAs have long been viewed as highly stable molecules, miRNA degradation pathways have now been identified in plants and algae, as revealed by Heriberto Cerutti and Fadia Ibrahim, and in animals, as discussed by us.

Even if this monograph cannot strive to be comprehensive in a field developing at such an amazing pace, I hope that the examples provided here will serve to illustrate the diversity of mechanisms regulating miRNAs, as well as highlight some unifying themes, particularly among the mechanisms regulating miRNA biogenesis. Undoubtedly, many more examples of regulation of miRNAs remain to be discovered and mechanistic details on known pathways to be revealed, promising an exciting future to this field of research.

Helge Großhans, PhD
Friedrich Miescher Institute for Biomedical Research
Basel, Switzerland

ABOUT THE EDITOR...

HELGE GROßHANS, PhD is a research group leader at the Friedrich Miescher Institute for Biomedical Research (FMI), which is part of the Novartis Research Foundation in Basel, Switzerland. His main research interests are in the developmental function, regulation, and mechanism of action of microRNAs. He received his PhD from the University of Heidelberg, Germany and did his postdoctoral training at Yale University, USA. He is the winner of a 2009 ERC Award.

PARTICIPANTS

Paola Briata
Istituto Nazionale per la Ricerca
 sul Cancro (IST)
Genova
Italy

Javier F. Cáceres
Medical Research Council
Human Genetics Unit
Institute of Genetics
 and Molecular Medicine
Western General Hospital
Edinburgh
UK

Heriberto Cerutti
School of Biological Sciences
Center for Plant Science Innovation
University of Nebraska
Lincoln, Nebraska
USA

Saibal Chatterjee
Friedrich Miescher Institute
 for Biomedical Research
Basel
Switzerland

Elisa Cuevas
Center for Anatomy
Institute of Cell Biology and Neurobiology
Charité-Universitätsmedizin Berlin
Berlin
Germany

Brandi N. Davis
Department of Biochemistry
Tufts University School of Medicine
and
Molecular Cardiology Research Institute
Tufts Medical Center
Boston, Massachusetts
USA

Witold Filipowicz
Friedrich Miescher Institute
 for Biomedical Research
Basel
Switzerland

Eleonora Franzoni
Center for Anatomy
Institute of Cell Biology and Neurobiology
Charité-Universitätsmedizin Berlin
Berlin
Germany

Sally Fujiyama-Nakamura
Institute of Molecular and Cellular
 Biosciences
University of Tokyo
Tokyo
Japan

Roberto Gherzi
Istituto Nazionale per la Ricerca
    sul Cancro (IST)
Genova
Italy

Richard I. Gregory
Stem Cell Program
Children's Hospital Boston
and
Department of Biological Chemistry
    and Molecular Pharmacology
Harvard Medical School
Harvard Stem Cell Institute
Boston, Massachusetts
USA

Helge Großhans
Friedrich Miescher Institute
    for Biomedical Research
Basel
Switzerland

Sonia Guil
Cancer Epigenetics and Biology Program
    (PEBC)
Catalan Institute of Oncology
    (ICO-IDIBELL)
L'Hospitalet Barcelona
Catalonia
Spain

Akiko Hata
Department of Biochemistry
Tufts University School of Medicine
and
Molecular Cardiology Research Institute
Tufts Medical Center
Boston, Massachusetts
USA

Bret S.E. Heale
Medial Research Council
Human Genetics Unit
Institute of Genetics and Molecular
    Medicine
Western General Hospital
Edinburgh
UK

Fadia Ibrahim
School of Biological Sciences
Center for Plant Science Innovation
University of Nebraska
Lincoln, Nebraska
USA

Shigeaki Kato
Institute of Molecular and Cellular
    Biosciences
University of Tokyo
Tokyo
Japan

Liam P. Keegan
Medial Research Council
Human Genetics Unit
Institute of Genetics and Molecular
    Medicine
Western General Hospital
Edinburgh
UK

René F. Ketting
Hubrecht Institute-KNAW
University Medical Centre Utrecht
Utrecht
The Netherlands

Nicolas J. Lehrbach
Wellcome Trust Cancer Research
Gurdon Institute and Department
    of Biochemistry
University of Cambridge
Cambridge
UK

Nicole-Claudia Meisner
Novartis Institutes for Biomedical Research
Novartis Campus
Basel
Switzerland

Eric A. Miska
Wellcome Trust Cancer Research
Gurdon Institute and Department
    of Biochemistry
University of Cambridge
Cambridge
UK

Gracjan Michlewski
Medical Research Council
Human Genetics Unit
Institute of Genetics and Molecular
    Medicine
Western General Hospital
Edinburgh
UK

Mary A. O'Connell
Medial Research Council
Human Genetics Unit
Institute of Genetics and Molecular
    Medicine
Western General Hospital
Edinburgh
UK

Andres Ramos
Division of Molecular Structure
National Institute for Medical Research
The Ridgeway, London
UK

Michael G. Rosenfeld
Howard Hughes Medical Institute
Department and School of Medicine
University of California, San Diego
La Jolla, California
USA

Agnieszka Rybak
Max Delbrück Center for Molecular
    Medicine
Berlin
Germany

Michele Trabucchi
Howard Hughes Medical Institute
Department and School of Medicine
University of California, San Diego
La Jolla, California
USA

Robinson Triboulet
Stem Cell Program
Children's Hospital Boston
and
Department of Biological Chemistry
    and Molecular Pharmacology
Harvard Medical School
Harvard Stem Cell Institute
Boston, Massachusetts
USA

F. Gregory Wulczyn
Center for Anatomy
Institute of Cell Biology and Neurobiology
Charité-Universitätsmedizin Berlin
Berlin
Germany

Kaoru Yamagata
Institute of Molecular and Cellular
    Biosciences
University of Tokyo
Tokyo
Japan

CONTENTS

1. microRNA BIOGENESIS AND FUNCTION:
        AN OVERVIEW 1

René F. Ketting

Abstract... 1
Introduction: PTGS in Plants and Small RNAs.. 1
RNAi.. 2
Dicer 2
Argonaute 4
The First microRNAs and Links to RNAi 4
miRNAs: Ancient Mediators of Gene Regulation 5
Primary miRNA Processing by Drosha 6
Small RNA Selectivity of Argonaute Proteins 7
Target Recognition.. 7
Mechanisms of miRNA-Mediated Silencing.. 8
Regulating miRNAs 10
Conclusion 10

2. REGULATION OF pri-miRNA PROCESSING THROUGH Smads................15

Akiko Hata and Brandi N. Davis

Abstract.. 15
Introduction: Basic TGFβ Signaling 15
miRNA Biogenesis.. 17
The First Processing Step by Drosha 19
R-Smads Regulate miRNA Maturation 20
Mechanism of Regulation of Specific pri-miRNAs by Smad 22
Transcriptional Regulation of miRNA Genes by Smads 24
Conclusion and Future Prospects.. 24

3. STIMULATION OF pri-miR-18a PROCESSING BY hnRNP A1 28

Gracjan Michlewski, Sonia Guil and Javier F. Cáceres

Abstract... 28
Introduction.. 28
hnRNP A1 Binds to The pri-miR-18a Stem-Loop Structure 29
hnRNP A1 Promotes The Drosha-Mediated Processing of pri-miR-18a 30
Role of The Terminal Loops in miRNA Processing 31
Conclusion and Future Prospects.. 33

4. KSRP PROMOTES THE MATURATION OF A GROUP OF miRNA
PRECURSORS 36

Michele Trabucchi, Paola Briata, Witold Filipowicz, Andres Ramos,
    Roberto Gherzi and Michael G. Rosenfeld

Abstract... 36
Introduction.. 36
Co-Activators and Co-Repressors of miRNA Precursor Maturation 37
Impact of KSRP and Other Co-Activators and Co-Repressors of miRNA
    Precursor Maturation on Cell Proliferation, Differentiation and Cancer.................. 39
Conclusion 41

5. HORMONAL REPRESSION OF miRNA BIOSYNTHESIS THROUGH
A NUCLEAR STEROID HORMONE RECEPTOR 43

Sally Fujiyama-Nakamura, Kaoru Yamagata and Shigeaki Kato

Abstract... 43
Introduction.. 43
p68/p72 DEAD-Box RNA Helicases Serve as RNA-Binding Components
    in the Drosha Complex 44
Gene Regulation by Nuclear Estrogen Receptors... 46
Estrogen-Induced mRNA Stability is Mediated through Hormonally Regulated
    miRNA Biosynthesis.. 49
Conclusion 52

6. AUTOREGULATORY MECHANISMS CONTROLLING
THE MICROPROCESSOR 56

Robinson Triboulet and Richard I. Gregory

Abstract... 56
Introduction.. 56
Posttranscriptional Regulation of DGCR8 by the Microprocessor............................... 59
Stabilization of Drosha Protein by DGCR8.. 61
Conclusion 61

7. REGULATION OF pre-miRNA PROCESSING 67

Nicolas J. Lehrbach and Eric A. Miska

Abstract.. 67
Introduction... 67
miRNAs and Developmental Timing in *C. elegans* 68
The Heterochronic Gene *lin-28* encodes a Regulator of *let-7* microRNA Processing 69
Lin28/LIN-28 Promotes Uridylation and Degradation of Pre-let-7 71
Heterochronic Gene Orthologues: Ancient Stem Cell Regulators? 71
Conclusion 73

8. THE EFFECT OF RNA EDITING AND ADARs ON miRNA BIOGENESIS AND FUNCTION 76

Bret S.E. Heale, Liam P. Keegan and Mary A. O'Connell

Abstract.. 76
Introduction... 76
Prevalence of Edited miRNAs... 78
Effects of Editing of pri-miRNAs and pre-miRNAs on Biogenesis 80
The Effect of Editing on miRNA Function 81
ADARs as Competing dsRNA-Binding Proteins 82
Editing of Seed Target Sequences within 3'UTRs.. 82
Conclusion 83

9. miRNAs NEED A TRIM: REGULATION OF miRNA ACTIVITY BY Trim-NHL PROTEINS 85

F. Gregory Wulczyn, Elisa Cuevas, Eleonora Franzoni and Agnieszka Rybak

Abstract.. 85
Introduction... 86
The Trim-NHL Family of Developmental Regulators....................................... 87
The Trim Domain as E3 Ubiquitin Ligase 89
Functional Analysis of Individual Trim-NHL Family Members 90
Trim-NHL Proteins as Regulators of the miRNA Pathway 95
Conclusion 99

10. PROPERTIES OF THE REGULATORY RNA-BINDING PROTEIN HuR AND ITS ROLE IN CONTROLLING miRNA REPRESSION......106

Nicole-Claudia Meisner and Witold Filipowicz

Abstract.. 106
Introduction to HuR and ARE Elements... 106
Regulation of the Regulator 108
Molecular Mechanisms of Posttranscriptional Control by HuR....................... 112
Function of HuR in the Relief of miRNA-Mediated Repression 114
Synergism between HuR and let-7 in Translational Repression of c-Myc mRNA............. 117
Conclusion 118

11. TURNOVER OF MATURE miRNAs AND siRNAs IN PLANTS AND ALGAE 124

Heriberto Cerutti and Fadia Ibrahim

Abtract 124
Introduction 125
Small RNA Processing 126
Small RNA Modification by 2′-O-Methylation 126
Small RNA Loading and Activation of the RNA-Induced Silencing Complex 128
Mature Small RNA Degradation by Ribonucleases 130
Quality Control of Mature Small RNAs 133
Conclusion 134

12. MicroRNases AND THE REGULATED DEGRADATION OF MATURE ANIMAL miRNAs 140

Helge Großhans and Saibal Chatterjee

Abstract 140
Introduction 140
microRNA Biogenesis and Function 141
XRN-2, a Multifunctional Exoribonuclease 143
Degradation of *C. elegans* miRNAs by XRN-2 144
Is miRNA Turnover a Substrate-Specific Event? 148
A Function of XRN-2 in miRNA Turnover beyond *C. elegans*? 148
Half-Lives of miRNAs—Not all miRNAs are Equal 149
Target Availability Affects Release of miRNAs
        from AGO and their Subsequent Degradation 151
Conclusion 152

INDEX 157

CHAPTER 1

microRNA BIOGENESIS AND FUNCTION
An Overview

René F. Ketting*

Abstract: During the last decade of the 20th century a totally novel way of gene regulation was revealed. Findings that at first glance appeared freak features of plants or *C. elegans* turned out to be mechanistically related and deeply conserved throughout evolution. This important insight was primed by the landmark discovery of RNA interference, or RNAi, in 1998. This work started an entire novel field of research, now usually referred to as RNA silencing. The common denominator of the phenomena grouped in this field are small RNA molecules, often derived from double stranded RNA precursors, that in association with proteins of the so-called Argonaute family, are capable of directing a variety of effector complexes to cognate RNA and/or DNA molecules. One of these processes is now widely known as microRNA-mediated gene silencing and I will provide a partially historical framework of the many steps that have led to our current understanding of microRNA biogenesis and function. This chapter is meant to provide a general overview of the various processes involved. For a comprehensive description of current models, I refer interested readers to the reviews and primary literature references provided in this chapter and to the further contents of this book.

INTRODUCTION: PTGS IN PLANTS AND SMALL RNAs

In the early 90s a number of papers were published that revealed an activity in Tobacco and Petunia plants that was triggered by repetitive transgenic DNA and that resulted in the silencing of that DNA and any other DNA bearing significant homology to the trigger (cosuppression).[1,2] At least part of these phenomena acted downstream of transcription, through destabilization of mRNA and hence was named "Post-Transcriptional Gene Silencing" (PTGS). The molecular trigger of this phenomenon was not clear, although

*René F. Ketting—Hubrecht Institute-KNAW and University Medical Centre Utrecht, Uppsalalaan 8, 3584 CT Utrecht, The Netherlands. Email: r.ketting@hubrecht.eu

Regulation of microRNAs, edited by Helge Großhans.
©2010 Landes Bioscience and Springer Science+Business Media.

it was speculated that "aberrant" RNA or double stranded RNA (dsRNA) were good candidates for priming PTGS. Although aberrant RNAs still play an important role in many models on RNA-mediated silencing events in plants, for example as templates on which dsRNA is synthesized, we now know that dsRNA is indeed in most cases the primary trigger. Furthermore, in 1999 a landmark paper from David Baulcombe and colleagues identified small RNA molecules as potential "specificity determinants" in PTGS.[3] This hypothesis has turned out to be absolutely correct and the identification of this type of small RNA species helped to lay the basis for an outburst of research activity on RNA-based silencing processes in the years that followed.

RNAi

In 1998, double-stranded RNA (dsRNA) was first described as a very potent and specific agent for gene silencing in *C. elegans*. The term RNA interference was coined to refer to the described silencing effects, a term that is now usually abbreviated to RNAi.[4] This ground-breaking work, published by Craig Mello, Andrew Fire and their colleagues, was awarded the Nobel Prize in 2006. Mello and Fire noted that RNAi targets exonic regions in RNA and leads to decreased RNA levels, consistent with a model in which RNAi leads to sequence specific mRNA destabilization, as had been found for cosuppression in plants. It was also noted that dsRNA could very well be a trigger in plant cosuppression, since inverted repeat sequences had been described as very potent triggers of PTGS. Soon after this paper, RNAi-like processes were identified in numerous other systems.[5] Biochemical experiments in *Drosophila* started to reveal a mechanistic framework of RNAi,[6,7] while genetics in *C. elegans* was revealing endogenous functions for RNAi and genes required for it.[8,9] It appeared that RNAi could mechanistically be roughly divided into two steps: an initiation step and an effector step (Fig. 1).[10] In the initiation step small RNAs are generated from the dsRNA trigger; in the effector step these small RNAs guide an Argonaute protein-containing complex named RNA-Induced Silencing Complex (RISC) to cognate mRNAs. The realization that small RNA molecules (then named siRNAs, for short interfering RNAs), like those described by Baulcombe in Tobacco plants undergoing PTGS, rather than long dsRNA molecules provided the sequence specificity of the whole process,[7,11] enabled efficient RNAi also in mammalian cells.[12] This provided a highly efficient way to perform reverse genetic experiments in cell culture systems, a finding that has revolutionized research on mammalian cells.

DICER

Dicer, the enzyme that generates siRNAs from dsRNA, was identified in 2001.[13] This enzyme contains two RNase III active sites, a so-called PAZ domain (named after three proteins in which this domain was first recognized: Piwi, Argonaute and Zwille), a helicase domain and a dsRNA-binding domain. It binds to the ends of dsRNA substrates and introduces a staggered double-stranded break further along the dsRNA.[14] The catalytic activity is very characteristic and always leaves a 3'-hydroxyl group, a 5'-phosphate group and a two base overhang at the 3' end. The length of the small RNA generated can vary, but usually is between 20 and 25 bases. Within one organism, different Dicer

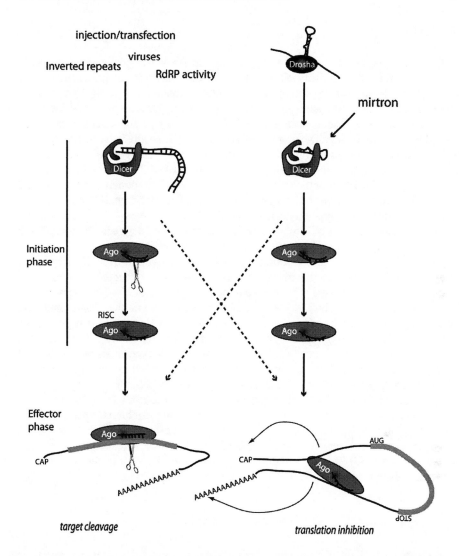

Figure 1. Schematic comparison between RNAi and miRNA mechanisms. For a more detailed scheme of miRNA action see Figure 2. "RdRP activity" refers to RNA-dependent RNA polymerase activity that in plants and yeast can turn ssRNA into dsRNA that is subsequently cleaved by Dicer. Likely, this is a major source for the dsRNA trigger in PTGS. The scissors indicate passenger strand and target cleavage. The dashed lines crossing from RNAi to miRNA and vice versa indicate that the separation between these pathways is not absolute: side effects from siRNAs in RNAi experiments can be triggered through miRNA like activities and miRNAs are capable of inducing target cleavage if presented with a properly matching target RNA. The type of silencing induced is also strongly dependent on the sub-type of Argonaute protein involved.

genes can be present, each encoding a protein generating rather specific subsets of small RNA products.[15] Mammals, however, only have one Dicer gene.

ARGONAUTE

Genetic experiments identified an Argonaute protein as a major player in RNAi in *C. elegans* soon after the discovery of RNAi itself.[16] Biochemical data supported this finding by identifying an Argonaute protein as an essential component of RISC.[17] Argonaute proteins had no known biochemical activities at that time. The only thing that was clear from Argonaute protein sequences is that they contained two characteristic domains: a PAZ domain (also found in Dicer) and a Piwi domain (named after the *Drosophila* Piwi protein, in which it was first recognized). It took a number of years before it became clear that Argonaute proteins actually form the catalytic center of RISC. The PAZ domain was shown to bind to the 3′ end of the small RNA,[18,19] while the Mid domain, in between the PAZ and Piwi domains, appeared to interact with the 5′ end of the small RNA.[20] Finally, the structure of the Piwi domain revealed an RNaseH-like structure,[21] consistent with the biochemical characteristics of RISC that had by then been well defined: endonucleolytic hydrolysis leaving a 5′-phosphate and a 3′-hydroxyl group.[22,23]

This RNA cleavage activity is first used for RISC activation: Initially Argonaute proteins are loaded with double-stranded siRNAs and in order to become active, one of the strands has to be removed. This can be done through endonucleolytic cleavage.[24-26] The discarded strand is referred to as the passenger strand, while the strand remaining bound to the Argonaute is known as the guide strand. It is this strand that guides the Argonaute protein to a target. At the target the very same catalytic activity used for RISC activation now can induce target RNA destabilization. However, this scenario is most likely an oversimplification, as there are indications that the catalytic activity of Argonaute proteins on small RNA duplexes differs from that on target RNA.[27,28]

It should also be noted that many Argonaute proteins contain a Piwi domain that is not compatible with nucleolytic activity. This has implications for both the mechanism of RISC activation as well as for the mechanism through which the targeted RNA is silenced. Passenger strand displacement in these Argonautes depends on weakened basepairing interactions within the small RNA duplex,[29] while target RNA silencing depends on additional cofactors recruited by the Argonaute (see below).

THE FIRST microRNAs AND LINKS TO RNAi

Already in 1993, years before the discovery of RNAi and siRNAs, two papers were published by the Ambros and Ruvkun labs describing a small RNA molecule in *C. elegans* that turned out to be the first microRNA ever to be described: *lin-4*.[30,31] *lin-4* was named as such because mutants display *lin*eage defects during development. The molecular basis of the lineage defects was the capability of *lin-4* to repress the activity of another gene, named *lin-14*, through imperfect basepairing interactions with the 3′UTR of the *lin-14* mRNA. It was also clear that *lin-4* came in two forms: a small 22 nucleotide version and a longer 61 nucleotide version that could fold into an imperfect hairpin structure. The small form of *lin-4* contained all the bases required for the basepairing interaction with *lin-14* and hence was likely the active, or mature form. In a follow-up study it was proposed that *lin-4* represses *lin-14* at the translational level.[32] The broader relevance of these findings remained unclear until 2000 when another small RNA gene was cloned: *let-7*.[33] This second small RNA had many features in common with *lin-4* but, in contrast to *lin-4*, turned out to be extremely well conserved across bilaterian animals.[34] This sparked the idea that gene regulation through small RNAs could be a much more widespread phenomenon than was appreciated at the time. Ruvkun and colleagues also proposed that the biogenesis and/or function of *lin-4* and *let-7*-like small RNAs could relate to the phenomenon of RNAi for which many biochemical studies had by then also indicated that small RNAs were providing the sequence specificity of the process (see above).

This notion appeared correct, as in 2001 a number of papers demonstrated that Dicer mediates both RNAi and *let-7* function[35-38] by processing the approximately 70 bases long *let-7* precursor into the mature small RNA and the longer dsRNA molecules used in RNAi into siRNAs. What's more, specific Argonaute proteins were shown to be essential for *let-7* function and processing,[35] just like another type of Argonaute protein had been shown to be required for RNAi.[16] Finally, *let-7* was found to trigger an RNAi-like reaction on target RNAs to which *let-7* could basepair perfectly.[39] Together, these findings made a very strong case for intimate mechanistic connections and similarities between *lin-4*-and-*let-7*-mediated gene regulation on the one hand and RNAi on the other (also see Fig. 1).

miRNAs: ANCIENT MEDIATORS OF GENE REGULATION

Not long after the finding that *let-7* was evolutionarily conserved, a number of reports appeared describing numerous endogenous small RNA genes from different animals, including *C. elegans*, *Drosophila* and human.[40-42] Like *let-7*, some of these were evolutionarily well conserved. They also derived from potential double stranded RNA structures, much like *let-7* and *lin-4*, and later studies indeed verified that for most of these novel endogenous small RNAs Dicer is required for the conversion of a longer precursor RNA into a mature small RNA.[43] Apart from their huge scientific impact, these findings also started to illustrate the need for a unifying name for this type of regulatory small RNA molecules: microRNAs (or miRNAs).[44]

Given all these results, it was absolutely clear that mRNA silencing by small RNA molecules was no exceptional feature of either *C. elegans* or plants, but represented a widespread and likely ancient mode of gene regulation. Indeed, miRNAs in the animal kingdom date far back, as far as the very base of the metazoan tree, suggesting that miRNAs have played an important role in the evolution of all animal life on our planet.[45] Notably, the complexity of the encoded miRNA repertoire correlates with the apparent complexity of the animal, raising the possibility that miRNAs have played a role in the growing complexity of multi-cellular metazoans. As in the animal kingdom, very primitive members of the plant kingdom produce miRNAs, as for example in the single celled "green yeast" chlamydomonas,[46] suggesting that also during plant evolution miRNAs have been around from very early-on. Interestingly however, while both plants and animals have a well-developed miRNA system, fungi do not appear to contain an equivalent of miRNA-mediated gene silencing, although a basic RNAi machinery is often present. This, in combination with the fact that plant and animal miRNA biogenesis and silencing mechanism appeared quite distinct in the first instance has been taken as an indication that miRNA-like pathways have independently evolved from the basic RNAi machinery in plants and animals.[47] However, recent discoveries on the mechanism of miRNA action in plants have started to unveil many more similarities between plant and animal miRNA systems, suggesting that despite the many differences, miRNA-type silencing may have started to evolve from the basic machinery even before plant and animal lineages split.[48]

Table 1. Factors involved in the various stages of miRNA biogenesis and function

	Arabidopsis	Drosophila	*C. elegans*	Mammals
First miRNA cleavage	DCL1	Drosha	DRSH-1	Drosha
	HYL1	Pasha	PASH-1	DGCR8
Second miRNA cleavage	DCL1	Dicer-1	Dcr-1	Dicer
	?	Loqs	?	TRBP, PACT
Nuclear export	HASTY	Exportin-5	?	Exportin-5
miRNA Argonaute	Ago1, Ago10	Ago-1	ALG-1, ALG-2	Ago1, 2, 3, 4
Silencing effector	?	GW182	AIN-1, AIN-2	TNRC6
Decapping	Varicose	Ge-1	?	EDC4
	?	Dcp-1, 2	?	Dcp1, 2
Deadenylation	?	?	?	Pan2-Pan3
	?	CAF1-NOT1	?	Ccr4-Caf1
	?	?	?	PABP

PRIMARY miRNA PROCESSING BY DROSHA

Already from the first description of *lin-4* it was clear that miRNAs are derived from RNA hairpin structures formed by RNAs of roughly 65-75 bases in length. As this hairpin RNA is the direct precursor for a mature miRNA it is now referred to as the pre-miRNA. However, it soon became clear that pre-miRNAs are only an intermediate between a long primary RNA transcript (known as pri-miRNA) and the mature form.[49] Pri-miRNAs are generated through transcription by RNA polymerase II and in most cases resemble regular genic transcripts in that they are capped, poly-A tailed and spliced.[50,51] In fact, many pre-miRNAs are generated from protein-coding loci, in which cases they are usually embedded in intronic sequences.[52]

While in Arabidopsis DCL1, one of the four available Dicer-like proteins, is the only RNaseIII-type nuclease required for the generation of most miRNA duplexes[53] and the whole process takes place in the nucleus, in animal systems Dicer only processes the pre-miRNA into the mature form and does so in the cytoplasm. The step from pri- to pre-miRNA occurs in the nucleus and is catalyzed by another RNase III-type enzyme, named Drosha[54] (also see Table 1). This enzyme, together with the RNA-binding protein Pasha, or DGCR8, and other components that together are named the microprocessor,[55-57] binds to the open-ended region of the miRNA-containing hairpin.[58] It then releases the pre-miRNA from the pri-miRNA by a double strand cleavage, leaving a two-base 3' single-stranded overhang. After transport to the cytoplasm, mediated by a specialized nuclear export factor named exportin 5,[59-61] Dicer recognizes the 3' overhang on the pre-miRNA through its PAZ domain, after which it cleaves off the loop, generating again a two-basepair 3' overhang. This product is then loaded into an Argonaute protein.

A minority of animal miRNAs can be processed without involvement of Drosha. In these cases, pre-miRNAs are directly derived from spliced introns, which are processed by a lariat-debranching enzyme into a suitable Dicer substrate. These microRNAs go by the name mirtrons.[62-64]

SMALL RNA SELECTIVITY OF ARGONAUTE PROTEINS

In most species, multiple Argonaute proteins are present and in many cases, specific Argonaute proteins have a strong preference for specific types of short RNA molecules. Two well-described determinants of this effect are the identity of the most 5' base of the future guide strand[65] and the thermodynamic properties of the small RNA duplex.[66,67] Regarding miRNAs in mammals however, these all seem to go into each of the four Argonaute proteins known as Ago1, 2, 3 and 4[68] and functional differences between these four proteins regarding miRNA function has so far remained unclear. Only Ago2 has a unique role in target mRNA cleavage during RNAi in mammalian cells.

Argonaute loading occurs in the context of the so-called RISC-Loading Complex, or RLC.[69-71] In general, this complex consists of a Dicer protein, an Argonaute and an RNA-binding protein. The human version of this complex consists of three subunits: Dicer, Argonaute and the RNA-binding protein TRBP.[72] In addition, this core complex can interact with additional RNA-binding proteins, such as the TRBP-related protein PACT.[73] Structural studies have begun to reveal how the core Dicer-Ago2-TRBP complex may enable the transition of a processed double-stranded siRNA from Dicer to Argonaute,[74] although no data at atomic resolution is yet available for the RLC.

Apart from a molecular understanding of how siRNAs can be transferred between two proteins, such structural studies will likely shed light on another intriguing observation regarding the maturation of miRNA:Argonaute complexes. This relates to the finding that for most miRNAs, one of the two strands is found in vast excess over the other. The dominating strand is named "mature miRNA" while the other strand is usually called the "miRNA star" (miR*) strand. Interestingly, this is not unique to endogenous miRNAs, as clear directional strand loading can be observed with synthetic, fully basepaired siRNA duplexes where the chance of each strand to become either a guide or a passenger strand is often not random. The physical basis behind this observation has been shown to relate to the thermodynamic properties of the basepairing in the small RNA duplex: in thermodynamically asymmetric duplexes the strand whose 5' end is basepaired least strongly will remain associated with the mature Argonaute complex (e.g., will become the mature miRNA or the siRNA guide strand).[75,76] The other strand is usually discarded and degraded, although so far no study has directly addressed the fate of the discarded strand in detail.

TARGET RECOGNITION

As described above, miRNAs guide the Ago protein to homologous mRNAs. However, not the complete sequence of the miRNA is relevant in this step: a region in the 5' part of the miRNA is the main determinant when it comes to target recognition by miRNAs.[77-81] This region, spanning from base two to seven or eight is also known as the "seed" region and structural studies have shown that precisely these bases are projected away from the Argonaute protein allowing efficient basepairing to other RNA molecules.[82-84] Surprisingly, in animal systems, this region alone is sufficient for regulation of the targeted RNA molecules, although more extensive basepairing of the target to regions more 3' in the miRNA can have an effect on target recognition.[79]

Although mRNAs can in principle be recognized along their complete length, meaning either in their 5' untranslated region (UTR), coding region or 3'UTR, most naturally occuring miRNA regulatory sites in animals have so far been found in the 3'UTR.[81] This may relate to their mode of action, which often does not include direct target cleavage but rather relates to processes affecting translation (see below), as target RNA cleavage by extensively base-paring miRNAs, which is a common event in plants, often occurs within the coding region.[85]

Given the small amount of sequence information used in target recognition by miRNAs and the fact that target site accessibility also plays an important role in the interaction between RISC and target RNA,[86] in silico target predictions are not easy to make. Just searching for so-called "seed-matches" results in enormously long lists of potential miRNA targets and more sophisticated algorithms had to be developed to generate more meaningful outputs. These algorithms for example take into account the conserved nature of a predicted miRNA target site and apply scores to certain base compositions directly flanking the miRNA-binding site and/or the complete secondary structure of the 3'UTR. Many of these programs produce very useful information to people studying individual miRNAs or protein-coding genes, but one should keep in mind that many of the applied rules are based on correlations and not on mechanistic studies. The output of these programs is therefore highly enriched in genuine miRNA targets, but at the gene-by-gene level one still has to take good care and validate predicted interactions through experiments. For a more in-depth review on this topic please see.[81]

MECHANISMS OF miRNA-MEDIATED SILENCING

As described above, some miRNAs regulate their mRNA target through direct target cleavage, analogous to the mRNA degrading activity observed during RNAi. As already described above, this cleavage event is a very specific one: nucleolytically active Argonaute proteins cleave targeted RNA molecules between the bases that pair to bases 10 and 11 of the small RNA, as counted from its 5' end,[11] leaving a 5' phosphate and a 3' hydroxyl group on the cleavage products.[22,23] These signatures have been used to demonstrate direct target RNA cleavage induced by miRNAs, as it is quite distinct from any other indirect RNA decay pathway. In order to facilitate such direct, Argonaute-mediated cleavage events, the miRNA requires significant basepairing beyond the seed region, including perfect basepairing in the region surrounding the site to be cleaved, i.e., bases 9-11 of the miRNA. This type of regulation is mainly found in plants, although recent work has shown that in plants translational inhibition, as discussed below, makes a major contribution to miRNA-mediated silencing in plants as well.[87,88]

In cases where miRNAs do not induce direct target cleavage, they lead to translational inhibition and mRNA destabilization through mechanisms other than Argonaute-mediated cleavage. The mechanism of translational inhibition by miRNAs has been and still is heavily debated with various models being proposed, ranging from effects directly on the initiation step, shortening of the polyA tail, effects on translation elongation and proteolysis of the nascent poly-peptide chain.[89] However, without going into the many details that can be discussed regarding this topic, many laboratories seem to converge on a model in which miRNAs interfere in some way with the initiation phase of translation, through recognition of the initiation complex or the cap structure[90-92] and in which the Argonaute protein needs to interacts with a GW-repeat-containing protein named GW182,

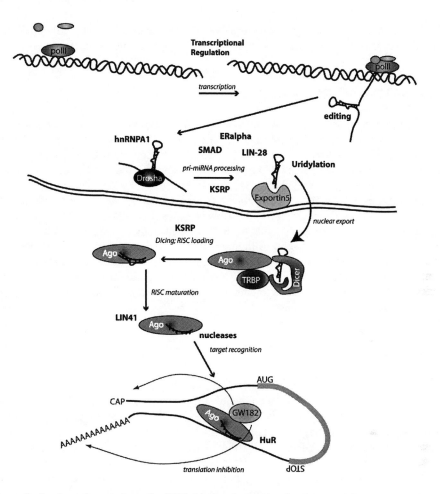

Figure 2. A schematic overview of miRNA biogenesis and function is presented. Important events are indicated in italics. Regulatory activities and/or factors are indicated in bold. The various factors/ activities are described in detail in the main chapters of this book.

also known as TNRC6, to achieve this.[93-97] Interestingly, miRNA-targeted mRNAs appear to be separated from the bulk of the active mRNA pool by sequestration into cytoplasmic foci, named P-bodies.[98] While initially P-bodies were considered mechanistically related to the miRNA silencing event, it has now been clearly shown that these structures do not to play an active role in the translational silencing process,[99] but rather might function as a storage place from which silenced mRNA can sometimes be reactivated.[100]

While none of the above interactions directly degrade the targeted mRNA, miRNA-mediated silencing often correlates with decreases in mRNA abundance.[80] Most likely this is the result of mRNA de-capping and de-adenylation processes, known to be associated with P-bodies and these processes do contribute to the overall silencing response,[87,95,101-105] suggesting that while P-body localization is not the main trigger in miRNA silencing, it may still be relevant to the total silencing effect imposed by miRNAs. In fact, large-scale proteomics experiments seem to indicate that mRNA destabilization may be even more important for general miRNA silencing than effects on translation.[106,107] Clearly, the way miRNAs impose their effects is still open to much debate.

REGULATING miRNAs

Many recent reports have illustrated that miRNA activity itself is subject to regulation. The main focus of the following chapters in this book will discuss in detail a number of mechanisms affecting miRNA-mediated gene regulation. Consequently, I limit myself to the notion that already at this point in time almost all levels in the miRNA pathway have been found to be subject to regulation: miRNA transcription, Drosha, Dicer, Argonaute, 3'UTR and miRNA molecules themselves are all affected by the action of additional proteins (Fig. 2). Furthermore, miRNA pathway genes themselves tend to be strongly regulated by miRNAs, suggesting that negative feed-back loops are essential for proper miRNA-mediated gene regulation. Although it may not be surprising that such a general gene regulatory mechanism is not allowed to operate unchecked, these findings emphasize that the rather simplistic way in which miRNA-mediated gene regulation has so far been approached will surely not suffice to completely understand the how, when and why of miRNA-mediated gene silencing.

CONCLUSION

In a very short time, the research field of small RNAs has grown enormously. The fact that a book can now be written on the mere regulation of just one of the small RNA pathways, the miRNA pathway, is by itself already a clear illustration of that notion. Less than 10 years after the realization that miRNAs represent a significant and widely conserved way of gene regulation, we know many of the core players (Table I) and we are starting to develop a sense of how this small RNA machinery is interwoven with the rest of the cell's processes (Fig. 2 and further chapters). Given the rapid pace of discovery of regulatory steps in the miRNA pathway at present, it seems inevitable that the findings described in this book will soon turn out to describe only our first encounters with miRNA regulatory steps. It appears likely that the miRNA pathway is manipulated in many different ways and a good understanding of these processes will be of great value not only for understanding miRNA function but also for a better appreciation for the many noncoding RNA molecules, other than RNAi-related small RNAs, that are presently being identified. Personally, I look forward to the many discoveries to be made in these fields in the coming years.

REFERENCES

1. Jorgensen RA, Atkinson RG, Forster RL et al. An RNA-based information superhighway in plants. Science 279 1998; 5356:1486-1487.
2. Kooter JM, Matzke MA, Meyer P et al. Listening to the silent genes: transgene silencing, gene regulation and pathogen control. Trends Plant Sci 4 1999; 9:340-347.
3. Hamilton AJ, Baulcombe DC. A species of small antisense RNA in post-transcriptional gene silencing in plants. Science 286 1999; 5441:950-952.

4. Fire A, Xu S, Montgomery MK et al. Potent and specific genetic interference by double-stranded RNA in Caenorhabditis elegans. Nature 391 1998; 6669:806-811.
5. Hutvagner G, Zamore PD. RNAi: nature abhors a double-strand. Curr Opin Genet Dev 12 2002; 2:225-232.
6. Hammond SM, Bernstein E, Beach D et al. An RNA-directed nuclease mediates post-transcriptional gene silencing in Drosophila cells. Nature 404 2000; 6775:293-296.
7. Zamore PD, Tuschl T, Sharp PA et al. RNAi: double-stranded RNA directs the ATP-dependent cleavage of mRNA at 21 to 23 nucleotide intervals. Cell 101 2000; 1:25-33.
8. Tabara H, Grishok A, Mello CC et al. RNAi in C. elegans: soaking in the genome sequence. Science 282 1998; 5388:430-431.
9. Ketting RF, Haverkamp TH, van Luenen HG et al. Mut-7 of C. elegans, required for transposon silencing and RNA interference, is a homolog of Werner syndrome helicase and RNaseD. Cell 99 1999; 2:133-141.
10. Hannon GJ. RNA interference. Nature 418 2002; 6894:244-251.
11. Elbashir SM, Lendeckel W, Tuschl T et al. RNA interference is mediated by 21- and 22-nucleotide RNAs. Genes Dev 15 2001; 2:188-200.
12. Elbashir SM, Harborth J, Lendeckel W et al. Duplexes of 21-nucleotide RNAs mediate RNA interference in cultured mammalian cells. Nature 411 2001; 6836:494-498.
13. Bernstein E, Caudy AA, Hammond SM et al. Role for a bidentate ribonuclease in the initiation step of RNA interference. Nature 409 2001; 6818:363-366.
14. Zhang H, Kolb FA, Jaskiewicz L et al. Single processing center models for human Dicer and bacterial RNase III. Cell 118 2004; 1:57-68.
15. Ramachandran V, Chen X. Small RNA metabolism in Arabidopsis. Trends Plant Sci 13 2008; 7:368-374.
16. Tabara H, Sarkissian M, Kelly WG et al. The rde-1 gene, RNA interference and transposon silencing in C. elegans. Cell 99 1999; 2:123-132.
17. Hammond SM, Boettcher S, Caudy AA et al. Argonaute2, a link between genetic and biochemical analyses of RNAi. Science 293 2001; 5532:1146-1150.
18. Ma JB, Ye K, Patel DJ et al. Structural basis for overhang-specific small interfering RNA recognition by the PAZ domain. Nature 429 2004; 6989:318-322.
19. Song JJ, Liu J, Tolia NH et al. The crystal structure of the Argonaute2 PAZ domain reveals an RNA binding motif in RNAi effector complexes. Nat Struct Biol 10 2003; 12:1026-1032.
20. Ma JB, Yuan YR, Meister G et al. Structural basis for 5'-end-specific recognition of guide RNA by the A. fulgidus Piwi protein. Nature 434 2005; 7033:666-670.
21. Song JJ, Smith SK, Hannon GJ et al. Crystal structure of Argonaute and its implications for RISC slicer activity. Science 305 2004; 5689:1434-1437.
22. Martinez J, Tuschl T. RISC is a 5' phosphomonoester-producing RNA endonuclease. Genes Dev 18 2004; 9:975-980.
23. Schwarz DS, Tomari Y, Zamore PD et al. The RNA-induced silencing complex is a Mg2+-dependent endonuclease. Curr Biol 14 2004; 9:787-791.
24. Matranga C, Tomari Y, Shin C et al. Passenger-strand cleavage facilitates assembly of siRNA into Ago2-containing RNAi enzyme complexes. Cell 123 2005; 4:607-620.
25. Leuschner PJ, Ameres SL, Kueng S et al. Cleavage of the siRNA passenger strand during RISC assembly in human cells. EMBO Rep 7 2006; 3:314-320.
26. Rand TA, Petersen S, Du F et al. Argonaute2 cleaves the anti-guide strand of siRNA during RISC activation. Cell 123 2005; 4:621-629.
27. Steiner FA, Okihara KL, Hoogstrate SW et al. RDE-1 slicer activity is required only for passenger-strand cleavage during RNAi in Caenorhabditis elegans. Nat Struct Mol Biol 16 2009; 2:207-211.
28. Wang B, Li S, Qi HH et al. Distinct passenger strand and mRNA cleavage activities of human Argonaute proteins. Nat Struct Mol Biol 2009.
29. Kawamata T, Seitz H, Tomari Y et al. Structural determinants of miRNAs for RISC loading and slicer-independent unwinding. Nat Struct Mol Biol 16 2009; 9:953-960.
30. Lee RC, Feinbaum RL, Ambros V et al. The C. elegans heterochronic gene lin-4 encodes small RNAs with antisense complementarity to lin-14. Cell 75 1993; 5:843-854.
31. Wightman B, Ha I, Ruvkun G et al. Post-transcriptional regulation of the heterochronic gene lin-14 by lin-4 mediates temporal pattern formation in C. elegans. Cell 75 1993; 5:855-862.
32. Olsen PH, Ambros V. The lin-4 regulatory RNA controls developmental timing in Caenorhabditis elegans by blocking LIN-14 protein synthesis after the initiation of translation. Dev Biol 216 1999; 2:671-680.
33. Reinhart BJ, Slack FJ, Basson M et al. The 21-nucleotide let-7 RNA regulates developmental timing in Caenorhabditis elegans. Nature 403 2000; 6772:901-906.
34. Pasquinelli AE, Reinhart BJ, Slack F et al. Conservation of the sequence and temporal expression of let-7 heterochronic regulatory RNA. Nature 408 2000; 6808:86-89.

35. Grishok A, Pasquinelli AE, Conte D et al. Genes and mechanisms related to RNA interference regulate expression of the small temporal RNAs that control C. elegans developmental timing. Cell 106 2001; 1:23-34.
36. Ketting RF, Fischer SE, Bernstein E et al. Dicer functions in RNA interference and in synthesis of small RNA involved in developmental timing in C. elegans. Genes Dev 15 2001; 20:2654-2659.
37. Hutvágner G, McLachlan J, Pasquinelli AE et al. A cellular function for the RNA-interference enzyme Dicer in the maturation of the let-7 small temporal RNA. Science 293 2001; 5531:834-838.
38. Knight SW, Bass BL. A role for the RNase III enzyme DCR-1 in RNA interference and germ line development in Caenorhabditis elegans. Science 293 2001; 5538:2269-2271.
39. Hutvagner G, Zamore PD. A microRNA in a multiple-turnover RNAi enzyme complex. Science 297 2002; 5589:2056-2060.
40. Lagos-Quintana M, Rauhut R, Lendeckel W et al. Identification of novel genes coding for small expressed RNAs. Science 294 2001; 5543:853-858.
41. Lau NC, Lim LP, Weinstein EG et al. An abundant class of tiny RNAs with probable regulatory roles in Caenorhabditis elegans. Science 294 2001; 5543:858-862.
42. Lee RC, Ambros V. An extensive class of small RNAs in Caenorhabditis elegans. Science 294 2001; 5543:862-864.
43. Ambros V. microRNAs: tiny regulators with great potential. Cell 107 2001; 7:823-826.
44. Ambros V, Bartel B, Bartel DP et al. A uniform system for microRNA annotation. RNA 9 2003; 3:277-279.
45. Grimson A, Srivastava M, Fahey B et al. Early origins and evolution of microRNAs and Piwi-interacting RNAs in animals. Nature 455 2008; 7217:1193-1197.
46. Molnár A, Schwach F, Studholme DJ et al. miRNAs control gene expression in the single-cell alga Chlamydomonas reinhardtii. Nature 447 2007; 7148:1126-1129.
47. Jones-Rhoades MW, Bartel DP, Bartel B et al. MicroRNAS and their regulatory roles in plants. Annu Rev Plant Biol 57 2006:19-53.
48. Voinnet, O, Origin, biogenesis and activity of plant microRNAs. Cell 136 2009; 4:669-687.
49. Lee Y, Jeon K, Lee JT et al. microRNA maturation: stepwise processing and subcellular localization. EMBO J 21 2002; 17:4663-4670.
50. Cai X, Hagedorn CH, Cullen BR et al. Human microRNAs are processed from capped, polyadenylated transcripts that can also function as mRNAs. RNA 10 2004; 12:1957-1966.
51. Lee Y, Kim M, Han J et al. microRNA genes are transcribed by RNA polymerase II. EMBO J 23 2004; 20:4051-4060.
52. Kim YK, Kim VN. Processing of intronic microRNAs. EMBO J 26 2007; 3:775-783.
53. Park W, Li J, Song R et al. CARPEL FACTORY, a Dicer homolog and HEN1, a novel protein, act in microRNA metabolism in Arabidopsis thaliana. Curr Biol 12 2002; 17:1484-1495.
54. Lee Y, Ahn C, Han J et al. The nuclear RNase III Drosha initiates microRNA processing. Nature 425 2003; 6956:415-419.
55. Han J, Lee Y, Yeom KH et al. The Drosha-DGCR8 complex in primary microRNA processing. Genes Dev 18 2004; 24:3016-3027.
56. Denli AM, Tops BB, Plasterk RH et al. Processing of primary microRNAs by the Microprocessor complex. Nature 432 2004; 7014:231-235.
57. Landthaler M, Yalcin A, Tuschl T et al. The human DiGeorge syndrome critical region gene 8 and Its D. melanogaster homolog are required for miRNA biogenesis. Curr Biol 14 2004; 23:2162-2167.
58. Han J, Lee Y, Yeom KH et al. Molecular basis for the recognition of primary microRNAs by the Drosha-DGCR8 complex. Cell 125 2006; 5:887-901.
59. Lund E, Güttinger S, Calado A et al. Nuclear export of microRNA precursors. Science 303 2004; 5654:95-98.
60. Yi R, Qin Y, Macara IG et al. Exportin-5 mediates the nuclear export of pre-microRNAs and short hairpin RNAs. Genes Dev 17 2003; 24:3011-3016.
61. Bohnsack MT, Czaplinski K, Gorlich D et al. Exportin 5 is a RanGTP-dependent dsRNA-binding protein that mediates nuclear export of pre-miRNAs. RNA 10 2004; 2:185-191.
62. Berezikov E, Chung WJ, Willis J et al. Mammalian mirtron genes. Mol Cell 28 2007; 2:328-336.
63. Okamura K, Hagen JW, Duan H et al. The mirtron pathway generates microRNA-class regulatory RNAs in Drosophila. Cell 130 2007; 1:89-100.
64. Ruby JG, Jan CH, Bartel DP et al. Intronic microRNA precursors that bypass Drosha processing. Nature 448 2007; 7149:83-86.
65. Mi S, Cai T, Hu Y et al. Sorting of small RNAs into Arabidopsis argonaute complexes is directed by the 5′ terminal nucleotide. Cell 133 2008; 1:116-127.
66. Steiner FA, Hoogstrate SW, Okihara KL et al. Structural features of small RNA precursors determine Argonaute loading in Caenorhabditis elegans. Nat Struct Mol Biol 14 2007; 10:927-933.

67. Tomari Y, Du T, Zamore PD et al. Sorting of Drosophila small silencing RNAs. Cell 130 2007; 2:299-308.
68. Farazi TA, Juranek SA, Tuschl T et al. The growing catalog of small RNAs and their association with distinct Argonaute/Piwi family members. Development 135 2008; 7:1201-1214.
69. Chendrimada TP, Gregory RI, Kumaraswamy E et al. TRBP recruits the Dicer complex to Ago2 for microRNA processing and gene silencing. Nature 436 2005; 7051:740-744.
70. Gregory RI, Chendrimada TP, Cooch N et al. Human RISC couples microRNA biogenesis and post-transcriptional gene silencing. Cell 123 2005; 4:631-640.
71. Tang G. siRNA and miRNA: an insight into RISCs. Trends Biochem Sci 30 2005; 2:106-114.
72. MacRae IJ, Ma E, Zhou M et al. In vitro reconstitution of the human RISC-loading complex. Proc Natl Acad Sci USA 105 2008; 2:512-517.
73. Lee Y, Hur I, Park SY et al. The role of PACT in the RNA silencing pathway. EMBO J 25 2006; 3:522-532.
74. Wang HW, Noland C, Siridechadilok B et al. Structural insights into RNA processing by the human RISC-loading complex. Nat Struct Mol Biol 16 2009; 11:1148-1153.
75. Schwarz DS, Hutvágner G, Du T et al. Asymmetry in the assembly of the RNAi enzyme complex. Cell 115 2003; 2:199-208.
76. Khvorova A, Reynolds A, Jayasena SD et al. Functional siRNAs and miRNAs exhibit strand bias. Cell 115 2003; 2:209-216.
77. Lewis BP, Shih IH, Jones-Rhoades MW et al. Prediction of mammalian microRNA targets. Cell 115 2003; 7:787-798.
78. Kloosterman WP, Wienholds E, Ketting RF et al. Substrate requirements for let-7 function in the developing zebrafish embryo. Nucleic Acids Res 32 2004; 21:6284-6291.
79. Brennecke J, Stark A, Russell RB et al. Principles of microRNA-target recognition. PLoS Biol 3 2005; 3:e85.
80. Lim LP, Lau NC, Garrett-Engele P et al. Microarray analysis shows that some microRNAs downregulate large numbers of target mRNAs. Nature 433 2005; 7027:769-773.
81. Bartel, DP. microRNAs: target recognition and regulatory functions. Cell 136 2009; 2:215-233.
82. Wang Y, Juranek S, Li H et al. Structure of an argonaute silencing complex with a seed-containing guide DNA and target RNA duplex. Nature 456 2008; 7224:921-926.
83. Wang Y, Sheng G, Juranek S et al. Structure of the guide-strand-containing argonaute silencing complex. Nature 456 2008; 7219:209-213.
84. Wang Y, Juranek S, Li H et al. Nucleation, propagation and cleavage of target RNAs in Ago silencing complexes. Nature 461 2009; 7265:754-761.
85. Jones-Rhoades MW, Bartel DP. Computational identification of plant microRNAs and their targets, including a stress-induced miRNA. Mol Cell 14 2004; 6:787-799.
86. Ameres SL, Martinez J, Schroeder R et al. Molecular basis for target RNA recognition and cleavage by human RISC. Cell 130 2007; 1:101-112.
87. Brodersen P, Sakvarelidze-Achard L, Bruun-Rasmussen M et al. Widespread translational inhibition by plant miRNAs and siRNAs. Science 320 2008; 5880:1185-1190.
88. Lanet E, Delannoy E, Sormani R et al. Biochemical evidence for translational repression by Arabidopsis microRNAs. Plant Cell 21 2009; 6:1762-1768.
89. Pillai RS, Bhattacharyya SN, Filipowicz W et al. Repression of protein synthesis by miRNAs: how many mechanisms? Trends Cell Biol 17 2007; 3:118-126.
90. Pillai RS, Bhattacharyya SN, Artus CG et al. Inhibition of translational initiation by Let-7 microRNA in human cells. Science 309 2005; 5740:1573-1576.
91. Thermann R, Hentze MW. Drosophila miR2 induces pseudo-polysomes and inhibits translation initiation. Nature 447 2007; 7146:875-878.
92. Zdanowicz A, Thermann R, Kowalska J et al. Drosophila miR2 primarily targets the m7GpppN cap structure for translational repression. Mol Cell 35 2009; 6:881-888.
93. Behm-Ansmant I, Rehwinkel J, Doerks T et al. mRNA degradation by miRNAs and GW182 requires both CCR4:NOT deadenylase and DCP1:DCP2 decapping complexes. Genes Dev 20 2006; 14:1885-1898.
94. Eulalio A, Huntzinger E, Izaurralde E et al. GW182 interaction with Argonaute is essential for miRNA-mediated translational repression and mRNA decay. Nat Struct Mol Biol 15 2008; 4:346-353.
95. Chen CY, Zheng D, Xia Z et al. Ago-TNRC6 triggers microRNA-mediated decay by promoting two deadenylation steps. Nat Struct Mol Biol 16 2009; 11:1160-1166.
96. Zipprich JT, Bhattacharyya S, Mathys H et al. Importance of the C-terminal domain of the human GW182 protein TNRC6C for translational repression. RNA 15 2009; 5:781-793.
97. Eulalio A, Helms S, Fritzsch C et al. A C-terminal silencing domain in GW182 is essential for miRNA function. RNA 15 2009; 6:1067-1077.

98. Liu J, Valencia-Sanchez MA, Hannon GJ et al. microRNA-dependent localization of targeted mRNAs to mammalian P-bodies. Nat Cell Biol 7 2005; 7:719-723.
99. Eulalio A, Behm-Ansmant I, Schweizer D et al. P-body formation is a consequence, not the cause, of RNA-mediated gene silencing. Mol Cell Biol 27 2007; 11:3970-3981.
100. Bhattacharyya SN, Habermacher R, Martine U et al. Relief of microRNA-mediated translational repression in human cells subjected to stress. Cell 125 2006; 6:1111-1124.
101. Eulalio A, Rehwinkel J, Stricker M et al. Target-specific requirements for enhancers of decapping in miRNA-mediated gene silencing. Genes Dev 21 2007; 20:2558-2570.
102. Fabian MR, Mathonnet G, Sundermeier T et al. Mammalian miRNA RISC recruits CAF1 and PABP to affect PABP-dependent deadenylation. Mol Cell 35 2009; 6:868-880.
103. Zekri L, Huntzinger E, Heimstädt S et al. The silencing domain of GW182 interacts with PABPC1 to promote translational repression and degradation of microRNA targets and is required for target release. Mol Cell Biol 29 2009; 23:6220-6231.
104. Giraldez AJ, Mishima Y, Rihel J et al. Zebrafish MiR-430 promotes deadenylation and clearance of maternal mRNAs. Science 312 2006; 5770:75-79.
105. Eulalio A, Huntzinger E, Nishihara T et al. Deadenylation is a widespread effect of miRNA regulation. RNA 15 2009; 1:21-32.
106. Baek D, Villén J, Shin C et al. The impact of microRNAs on protein output. Nature 455 2008; 7209:64-71.
107. Selbach M, Schwanhäusser B, Thierfelder N et al. Widespread changes in protein synthesis induced by microRNAs. Nature 455 2008; 7209:58-63.

CHAPTER 2

REGULATION OF pri-miRNA PROCESSING THROUGH Smads

Akiko Hata* and Brandi N. Davis

Abstract: microRNAs (miRNAs) are small (~22 nucleotides (nt)), noncoding RNAs that play a critical role in diverse biological functions by modulating mRNA stability and translational control. Numerous miRNA profiling studies have indicated that the levels of miRNAs are tightly controlled during developmental stages and various pathophysiological and physiological conditions. Following transcription, the long primary miRNA transcript undergoes a series of coordinated maturation steps to generate the mature miRNA. Signaling pathways that control miRNA biogenesis and the mechanisms of regulation, however, are not well understood. In this chapter, we will discuss the finding that signal transducers of the Transforming Growth Factor β (TGFβ) signaling pathway, the Smads, play a critical regulatory role in the nuclear processing of miRNAs by the RNase III-type protein Drosha.

INTRODUCTION: BASIC TGFβ SIGNALING

TGFβ signaling pathways are fundamental to metazoan development and adult tissue homeostasis and are involved in the regulation of a variety of processes including differentiation, proliferation and migration.[1] Deregulation of the pathways is implicated in various developmental defects and human diseases, including cancer and cardiovascular disorders.[2,3] The TGFβ-family of ligands, such as TGFβ, activins and Bone Morphogenetic Proteins (BMPs), transmit biological information to cells by binding to cell surface receptors (Fig. 1). Binding of the ligand triggers the formation of heteromeric receptor complexes composed of the Type I and Type II receptors[1,4] (Fig. 1). Both the Type I and the Type II receptors contain serine/threonine kinase domains. Upon formation of the heteromeric receptor complexes, the constitutively active Type II kinase phosphorylates the juxtamembrane region of the Type I receptor and turns on the Type I receptor kinase

*Corresponding Author: Akiko Hata—800 Washington Street, Box 8486, Boston, MA 02111, USA
  Email: akiko.hata@tufts.edu

Regulation of microRNAs, edited by Helge Großhans.
©2010 Landes Bioscience and Springer Science+Business Media.

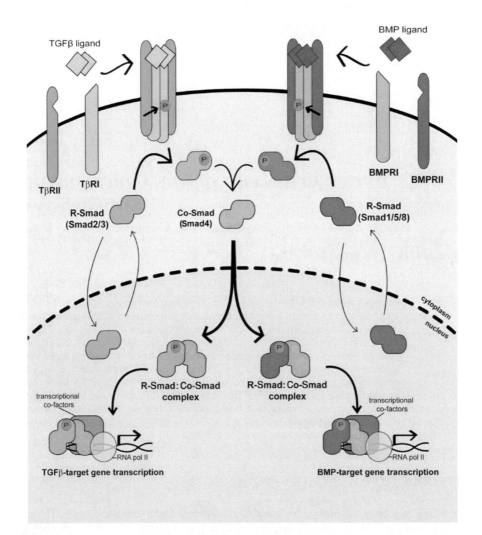

Figure 1. TGFβ and BMP signaling. TGFβs and BMPs signal by binding to Type II and Type I receptors, resulting in activation of the Ser/Thr kinase domain of the Type I receptor and subsequent phosphorylation of R-Smads. An activated R-Smad interacts with a Co-Smad generating a complex that translocates to the nucleus. In the nucleus, R-Smad and Co-Smad form a complex with the appropriate DNA-binding cofactor and together bind to a target gene promoter to induce or repress transcription.

activity. The activated Type I receptor kinase then phosphorylates cytoplasmic signal transducer proteins called receptor-activated Smad family proteins (R-Smads) at the carboxyl (C)-terminal serine residues (Fig. 1). The Type I receptors of TGFβs phosphorylate Smad2 and Smad3, while the Type I receptors of BMPs phosphorylate Smad1, Smad5 and Smad8. Upon phosphorylation, R-Smads associate with the common-partner Smad (Co-Smad), Smad4 and then translocate to the nucleus as an R-Smad-Smad4 complex.[1] In the nucleus, Smads are known to bind to a 5 base pair (bp) sequence (5′-CAGAC-3′) known as the Smad-binding element (SBE) through the conserved amino (N)-terminal Mad homology 1 (MH1) domain[5] (Fig. 1). The R-Smad-Co-Smad complex is known to interact with various transcription factors through the C-terminal MH2 domain, which increases the binding affinity and specificity of the Smad complex with target gene promoters. Upon binding to DNA, Smads and their transcription partners recruit transcription activators, such as histone acetyltransferases p300 or C/EBP-binding protein (CBP) and induce transcription of some target genes. Alternatively, in some circumstances Smads recruit transcription repressors, such as histone deacetylases and SKI/SNON and repress transcription of other target genes.[5] Although some specific functions of the TGFβ signal are known to be transmitted in a Smad-independent manner, Smad proteins and their gene regulatory function play a fundamental role in the general functions of the TGFβ signaling pathway.[1] While Co-Smad has been known as an essential partner of R-Smads in DNA binding and transcriptional regulation,[6] several studies have reported that a subset of gene regulation events mediated by TGFβ signal are observed in Smad4-null cells,[7,8] suggesting that R-Smads might be able to modulate gene expression without Co-Smad at the level of transcription or through a novel mechanism.

miRNA BIOGENESIS

miRNAs have been reported to control diverse aspects of biology, including developmental timing, differentiation, proliferation, cell death and metabolism. At least 30% of human genes are thought to be regulated by miRNAs. Approximately 30-50% of miRNAs are encoded within the introns of protein coding genes while the remaining miRNAs are located in intergenic sites.[9] The majority of miRNAs are transcribed by RNA polymerase (RNA pol) II and bear the 5' 7-methyl guanylate cap and 3' poly (A) tail, characteristic of mRNAs[10,11] (Fig. 2). The evolutionarily conserved mechanism that gives rise to mature miRNA involves two sequential endonucleolytic cleavages by the RNase III enzymes Drosha and Dicer (Fig. 2). Following transcription by RNA pol II, Drosha processes the primary miRNA transcript (pri-miRNA) into a ~65-80 nt hairpin structure termed the precursor-miRNA (pre-miRNA). Through the interaction with exportin-5 and Ran-GTP, the pre-miRNA is transported into the cytoplasm, where it undergoes a second round of processing catalyzed by Dicer (Fig. 2). This cleavage event gives rise to a double-stranded ~22 nt product comprised of the mature miRNA guide strand and the miRNA* passenger strand. The miRNA guide strand is then loaded into the RNA-Induced Silencing Complex (RISC) while the passenger strand is degraded (Fig. 2). The RISC complex loaded with miRNA associates with target mRNAs which then leads, in a majority of cases, to negative regulation of protein synthesis or mRNA degradation. The association of miRNAs with target mRNAs requires the presence of binding sites that are partially complementary to the miRNA sequences. The 5' region of the miRNA (nt 2-7), termed the 'seed sequence' seems especially important for mRNA recognition and repression by miRNAs.[12,13] Functional miRNA-binding sequences are often located in the 3'-untranslated region (UTR) of the target mRNA, but can also occur within the 5'UTR[14] or coding region.[15] Because miRNAs exert a variety of physiological functions primarily through the repression of target genes, the determination of miRNA targets has been an area of intense research. Computational and experimental approaches indicate that a single miRNA may target several dozen or even hundreds of mRNA.[16,17] Although major progress has been made in understanding the fundamental mechanism of miRNA biogenesis, little is known about the mechanisms that regulate this process.

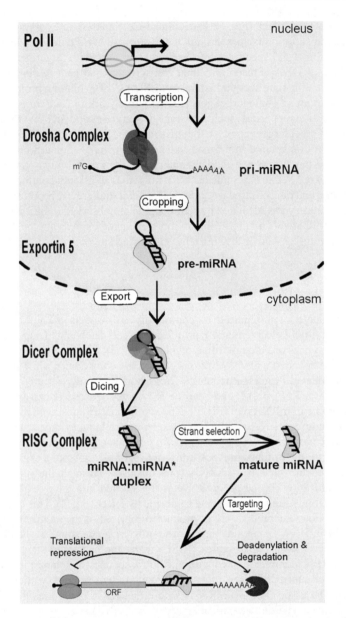

Figure 2. miRNA biogenesis pathway. Transcription by RNA pol II leads to capped and polyadenylated pri-miRNAs, which are processed by Drosha in the nucleus to generate pre-miRNAs. After translocation to the cytoplasm by exportin 5, pre-miRNAs are processed by Dicer to form mature miRNA/miRNA* duplexes. Following processing, miRNAs are assembled into the RISC complex. Only one strand of the duplex is stably associated with the RISC complex. The mature miRNA directs repression of mRNA containing partially complementary miRNA-binding sites within the 3′UTR.

Given the importance of miRNAs in development, it is not surprising that deregulation of miRNA expression is observed in a variety of developmental defects and human diseases, including cancer, neurological and cardiovascular disorders. A majority of miRNAs are located at sites of genomic instability, including duplications and fragile sites.[18,19] Global miRNA expression is frequently reduced in tumor samples relative to normal tissues, suggesting a role for miRNA in maintaining the differentiated state. For example, let-7 expression is often dramatically reduced in lung cancer and exogenous expression of let-7 can dramatically inhibit tumor growth in vivo.[20] Conversely, a subset of miRNAs, including miR-21 and miR-155, have been identified as highly expressed in a variety of tumors and may serve to promote tumor growth through the inhibition of pro-apoptotic pathways. miRNA expression profiling serves as a better predictor of tumor origin and prognosis than conventional gene arrays, further emphasizing the importance of miRNAs in oncogenesis.[21] Understanding the mechanisms that control both normal and deregulated miRNA expression may lead to new avenues for the treatment of a variety of disorders.

THE FIRST PROCESSING STEP BY DROSHA

The first step of miRNA maturation is cleavage of the pri-miRNAs at the stem of the hairpin structure by the nuclear RNase III enzyme Drosha. Cleavage by Drosha results in the release of the pre-miRNAs.[22] Although Drosha contains a highly conserved RNase III domain, purified Drosha is unable to specifically and efficiently generate pre-miRNAs in vitro.[23] This suggests that other cofactors may be required for Drosha activity. Indeed, gel filtration chromatography revealed that Drosha forms a large complex known as the "Microprocessor complex", including Drosha and an essential cofactor, DiGeorge syndrome critical region gene 8 protein (DGCR8, also known as Pasha) (Fig. 2), which can promote the efficient cleavage of pri-miRNA in vitro.[23] The pri-miRNA is composed of a ~33 nt stem connected by a terminal loop and flanked by single-stranded segments. DGCR8 is thought to recognize the region between the single-stranded RNA and the double-stranded stem to direct Drosha cleavage one helical turn (11bp) away from this junction.[24] While the cropping of many miRNAs can be mediated in vitro by purified DGCR8 and Drosha, nuclear run-on and in vitro processing assays indicate that the pri-miRNA to pre-miRNA cleavage of some miRNA is relatively slow and inefficient.[25] Therefore, the efficient processing of some miRNAs by the Drosha/DGCR8 complex may require the involvement of accessory factors. Furthermore, the precise position and orientation of Drosha cleavage serves a critical role in the generation of miRNAs as it determines the identity of the terminal nucleotide at either the 5' or the 3' end of the mature miRNA. An error in the Drosha cleavage site may result in the alteration of the miRNA seed sequence and cause redirection of miRNA targets. Additionally, in some cases, altered cleavage could invert the relative stability of the two miRNA strands, leading to the incorporation of the improper miRNA strand into the RISC complex. The association of Drosha accessory factors may promote the fidelity and activity of Drosha cleavage. Immunoprecipitation studies of tagged Drosha expressed in HEK293 cells suggest that at least 20 polypeptides may be associated with Drosha in vivo.[26] Furthermore, several recent reports suggest that cellular stimuli alters the association of Drosha with accessory factors.[27,28] The mechanisms which determine the precise composition of the Drosha complex in the processing of specific pri-miRNAs is unclear.

R-Smads REGULATE miRNA MATURATION

Recently, the DEAD-box RNA helicases p68 (DDX5) and p72 (DDX17) were identified as components of the large Drosha-mediated processing complex by immunopreciptition-mass spectrometry analysis and subsequently shown to also associate with DGCR8.[29,30] Members of the DEAD box family of proteins have been shown to act as RNA helicases, using the energy from ATP hydrolysis to unwind RNA structures or dissociate RNA-protein complexes. Additionally, p68 and p72 have been reported to interact with a variety of proteins, including transcription factors such as the Smads, p53 and estrogen receptor. Analysis of mature miRNA levels indicated reduced steady-state levels of miRNAs in p68(–/–) or p72(–/–) mouse embryonic fibroblasts (MEFs) in comparison with wild-type MEFs, suggesting an important role for p68 and p72 in miRNA biogenesis.[29] Although the exact role of p68 or p72 as subunits of the Drosha complex is unclear, it has been suggested that p68 and p72 are required to specifically recognize and stably bind to a subset of pri-miRNAs to initiate cleavage at precise sites by Drosha, similarly to DGCR8.[29] Interestingly, expression levels of a distinct set of, but not all, miRNAs are lowered in p68(–/–) MEFs and p72(–/–) MEFs, respectively and decreased expression of some of miRNAs are observed in both p68(–/–) MEFs or p72(–/–) MEFs,[29] suggesting that a subset of miRNAs are processed in p68- or p72-specific manner, while other miRNAs require both p68 and p72 for the processing.

The role of p68/p72 in miRNA processing is further supported by the positive regulation of Drosha-mediated processing mediated by the p68-interacting Smad proteins. The Smads are the signal transducers of the TGFβ family signaling cascade. TGFβ and its family member, BMP4 are particularly important for the differentiation of vascular smooth muscle cells (VSMCs). Treatment with either BMP4 or TGFβ increases expression of contractile smooth muscle genes. This process is due, at least in part, to the miR-21-mediated repression of programmed cell death protein-4 (PDCD4). miR-21 is rapidly induced by BMP4 and TGFβ in VSMC which results in a subsequent decrease in PDCD4 and increased VSMC gene expression.[27] Interestingly, although knockdown of the R-Smads prevents upregulation of mature and pre-miR-21 in response to BMP4 or TGFβ, no alteration in pri-miR-21 transcription is detected.[27] Furthermore, BMP4 or TGFβ could increase the expression of pre-miR-21 and mature miR-21 by facilitating the Drosha-mediated processing step. The identification of R-Smads as binding partners of p68 by yeast-two-hybrid suggested that R-Smads could associate with the Drosha complex.[31] Consistently, co-immunoprecipitation (co-IP) and RNA-IP studies confirmed that Smad is present in a complex with Drosha and p68 on the pri-miR-21 hairpin following BMP4 or TGFβ stimulation in vivo[27] (Fig. 3). Drosha binding to pri-miR-21 was also elevated following ligand treatment, suggesting that Smads may promote the association of Drosha with miRNA hairpins. These results indicate that TGFβ can regulate gene expression not only through direct transcriptional regulation but also through the regulation of miRNA processing (Fig. 3). Nucleo-cytoplasmic shuttling of Smads is tightly controlled by phosphorylation of serine residues at the C-terminus by the TGFβ Type I receptor kinases. Interestingly, mitogen-activated protein kinase (MAPK) and glycogen synthase kinase 3 (GSK3) can also alter the subcellular localization of Smads through phosphorylation in the linker region.[32,33] Thus, it is possible that Smad-dependent regulation of miRNA biosynthesis could be modulated independently of TGFβ and BMPs by signals that alter the nuclear localization of Smads, such as the ERK-MAPK and the Wnt pathways.

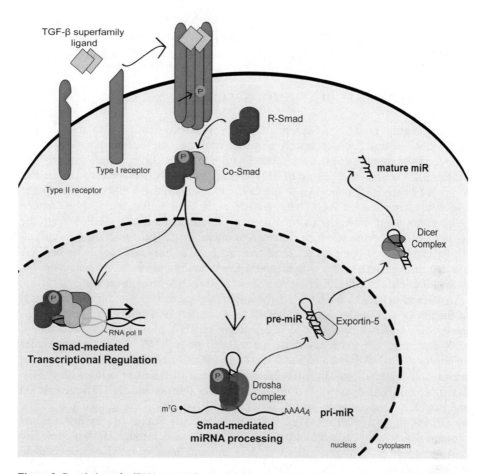

Figure 3. Regulation of miRNA maturation by TGFβ superfamily signaling. TGFβ and BMP signaling stimulates the production of pre-miR-21 by promoting the Drosha-mediated processing by controlling nuclear localization of R-Smad proteins. Thus, Smads regulate gene expression in two distinct manners; (i) transcriptional regulation by DNA binding and (ii) regulation of miRNA maturation by associating with the Drosha/DGCR8 complex.

Interestingly, the association of R-Smads with the Drosha-mediated processing machinery does not require the Co-Smad Smad4. Knockdown of *Smad4* in VSMC did not affect induction of miR-21; furthermore, miR-21 is strongly induced by TGFβ in the Smad4-null MDA-MB-468 breast cancer cell line.[27] It was previously reported that R-Smads and Smad4 translocate into the nucleus as a complex.[34] A more recent study, however, demonstrates that R-Smads and Smad4 can be independently transported into the nucleus through different nuclear import machineries.[35,36] Thus, R-Smads that are not locked into a complex with Smad4 might preferentially participate in miRNA processing through association with the Drosha/DGCR8 complex. In contrast, the R-Smad/Smad4 heteromeric complex may preferentially associate with the SBE in promoter regions of the TGFβ target genes and act as a transcription factor. miR-21 is highly expressed in a variety of tumors and implicated in tumor growth through the inhibition of pro-apoptotic pathways by targeting tumor suppressor genes. High levels of expression of miR-21 are associated with increased expression of TGFβ and implicated in poor prognosis in human breast cancer.[37]

MECHANISM OF REGULATION OF SPECIFIC pri-miRNAs BY Smad

To identify novel miRNAs regulated by R-Smads similarly to miR-21 and miR-199a,[27] a miRNA expression profiling analysis was performed using RNA samples from human VSMCs stimulated with BMP4 or TGFβ. Approximately 5% of the miRNAs analyzed (20 out of 377), including miR-21, were induced more than 1.6-fold by both BMP4 and TGFβ. Interestingly, many of these miRNAs contain the consensus sequence 5'-CAGAC-3', identical to the SBE, within the double-stranded stem region of pri-miRNAs (named R-SBE). R-Smads bind to R-SBE through the N-terminal MH1 domain which is known to interact with SBE in DNA. Similarly to the R-Smad-Smad4 complex which shows increased affinity and specificity of DNA binding through interaction with different DNA-binding proteins, we speculate that R-Smads might gain affinity and specificity of pri-miRNA-binding through interaction with p68 and/or Drosha. Mutations in the R-SBE abrogates TGFβ-induced recruitment of Smads as well as Drosha and DGCR8 to pri-miRNAs; additionally, mutation of the R-SBE impairs ligand-dependent pri- to pre-miRNA processing. Introduction of an R-SBE sequence to an otherwise unregulated pri-miRNA is sufficient to confer an ability to bind to Smad and hence the TGFβ/BMP -mediated regulation of pri- to pre-miRNA maturation. Thus, (i) direct association of Smad proteins with R-SBE in mature miRNAs operates as a molecular tag for Drosha and DGCR8 recognition and preferential association with a set of pri-miRNAs, facilitating their processing by Drosha upon TGFβ or BMP4 stimulation and (ii) Smads are multifunctional proteins which modulate gene expression transcriptionally through DNA binding and posttranscriptionally by pri-miRNA binding and regulation of miRNA processing in the nucleus.[27]

Recently, p53 and estrogen receptor α (ERα) were demonstrated to regulate processing by the Drosha microprocessor complex[28,38] (Fig. 4, and see chapter by Fujiyama-Nakamura et al). While p53 increases the expression of a subset of miRNAs by facilitating the processing of pri-miRNAs by the Drosha complex through interaction with the RNA helicase p68, similarly to Smads,[38] ERα bound to estradiol (E2) inhibits the production of a subset of miRNAs by attenuating the pri- to pre-miRNA processing through interaction with p68 and p72 and promoting a dissociation of the Drosha microprocessor complex from pri-miRNAs[28] (Fig. 4). miRNA expression profiling analysis in VSMCs includes a few miRNAs that are rapidly downregulated upon TGFβ or BMP4 stimulation (B.D. and A.H. unpublished observation). Together with a report that the expression of miR-206 is downregulated by BMP2 treatment at the post-transcriptional step in the myoblastic C2C12 cell line,[39] it is plausible to speculate that R-Smads might be able to regulate the Drosha microprocessor activity not only positively but also negatively, similarly to ERα.

It seems that only a subset of miRNAs are regulated by p53 or ERα, however, it is yet unclear what determines a selective regulation of specific miRNA processing by these proteins. It is possible that some levels of specificity are provided by the RNA helicases p68 or p72, with which regulatory proteins interact, as p68 and p72 seem

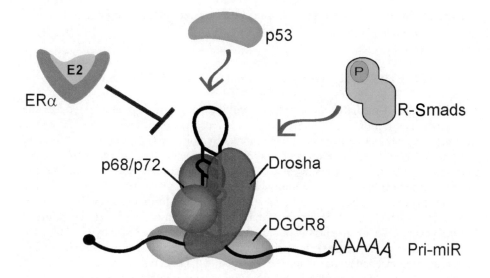

Figure 4. Positive and negative regulation of Drosha-mediated processing. ERα in the presence of estradiol (E2) blocks, while R-Smads and p53 positively regulate Drosha-mediated processing of a subset of p68/p72-dependent miRNAs.

to participate in the synthesis of partially overlapping but distinct sets of miRNAs.[29] Another possibility is that p53 or ERα bind directly to pri-miRNAs in sequence- or RNA secondary structure-specific manner. Based on evidence that p53 or ERα modulate different sets of miRNAs from Smad-regulated miRNAs, it is intriguing to speculate that p53 or ERα recognizes a specific RNA sequence or structure distinct from R-SBE. It is of note that p53 is known to bind both DNA and RNA,[40] thus p53 might play dual functions both in transcription and maturation of pri-miRNA, similarly to Smads. In addition to Smad and p53, other transcription factors known to bind both DNA and RNA include TFIIIA, Stat1 and WT1.[41] It was also reported that NFκB binds sequence-specifically to a DNA duplex and to a synthetic RNA aptamer predicted to form a stem-bulge-stem-loop structure with indistinguishable affinity and stoichiometry.[41] The RNA helicases p68 and p72 have been shown to interact with several additional DNA binding proteins, including MyoD,[42] Runx2[42] androgen receptor[43] and β-catenin.[44] Thus, it is intriguing to speculate that some of these DNA binding proteins might have dual roles and participate in the pri-miRNA processing as a part of the Drosha microprocessor complex.

An additional role of Smads in the regulation of the Drosha complex has been suggested by the study of a nuclear factor called Smad nuclear interacting protein 1 (SNIP1). SNIP1, which was originally identified as a nuclear protein partner of Smads, is found in a complex with Drosha.[45] The Arabidopsis homologue of SNIP1, DAWDLE (DDL), is required for efficient pri-miRNA to pre-miRNA processing and is thought to promote the access or recognition of pri-miRNA by the Arabidopsis homologue of Drosha; DCL1.[45] Furthermore, downregulation of SNIP1 in mammalian cells reduces the expression of subset of miRNAs, including miR-21.[45] These results suggest that

SNIP1 might participate in miRNA biogenesis by facilitating the Drosha function possibly through interaction with Smad proteins.

TRANSCRIPTIONAL REGULATION OF miRNA GENES BY SMADs

RNA pol II-mediated transcription provides a major regulatory step for the biosynthesis of miRNAs. A large-scale nucleosome positioning and chromatin immunoprecipitation-on-genomic DNA microarray chip (or ChIP-on-chip) analysis of the promoters of miRNA genes suggests that the promoter structure of miRNA genes, including the relative frequencies of CpG islands, transcription initiator elements and histone modifications, is indistinguishable between the promoters of miRNA genes and protein-coding genes.[46,47] Furthermore, DNA-binding factors that regulate miRNA transcription largely overlap with those that control protein-coding genes. Therefore, it is likely that the Smad proteins might modulate expression of miRNAs by regulating the transcription of miRNA genes (Fig. 2). Indeed, in the kidney, TGFβ activates AKT through the transcriptional induction of miRNAs targeting PTEN; miR-216a and miR-217.[48] Additionally, it is likely that the TGFβ signaling pathways might modulate transcription of genes encoding critical enzymes or regulators of miRNA biogenesis, including Drosha, DGCR8, Dicer, or Argonaute (Ago) proteins, which are required for formation of the RISC complex. Recently, the protein stability of Ago proteins was found to be regulated by posttranslational modification.[49] The relative level of Ago proteins may critically regulate the stability of miRNA as knockdown or overexpression of Agos markedly decreases or increases miRNA levels, respectively.[50] Ago2 protein stability was found to be regulated by hydroxylation of a specific proline residue (Pro 700) by the Type I collagen prolyl-4-hydroxylase [C-P4H(I)].[49] The Pro 700 is conserved among all four Ago proteins (Ago1-4) found in mammals, however, Ago2 and Ago4 seem to be hydroxylated to a greater extent than Ago1 and Ago3 in vivo.[49] A single amino acid mutation of Pro 700 to alanine of Ago2 reduced the steady-state expression of Ago2 protein.[49] Knockdown of the α or β subunit of C-P4H(I) in the cell reduced Ago2 protein stability, as well as RISC activity mediated by Ago2,[49] suggesting that this posttranslational modification might regulate the RNA interference (RNAi) mechanism via modulation of the protein stability of Ago2 and possibly other Agos.[49] The α subunit of C-P4H(I) is rate-limiting for the formation of active C-P4H(I) and has been reported to be transcriptionally induced by different stimuli, including TGFβ.[51] Thus, it is plausible that induction of C-P4H(I) by TGFβ may result in the stabilization of Ago proteins, which may result in general increase in RNAi activity and/or global increase in miRNA expression.

CONCLUSION AND FUTURE PROSPECTS

miRNAs are generated through the concerted action of multi-subunit complexes which promote the sequential cleavage, export and loading of miRNAs into RISC complexes. An increasing number of reports suggest that each of these steps serves as a potential point of regulation and therefore provides additional complexity to miRNA-dependent gene regulation. So far, there are only a few reports demonstrating mechanism of regulation of miRNA maturation. In comparison with traditional transcriptional role of Smads, which requires both transcription and translation steps in order to exhibit a change in protein expression, the novel function of Smads on miRNA biogenesis could more rapidly modify protein expression in response to ligand stimulation. Furthermore, it does not require the Co-Smad Smad4, which is frequently deleted in various tumors, such as pancreatic and colorectal tumors, thus gene regulation by Smad via miRNA is intact even in Smad4-null cells. Therefore, regulation of miRNA biogenesis may serve as the first line of response following TGFβ/BMP stimulation both in normal and tumor cells. Finally, as a single miRNA modulates the expression of hundreds of targets simultaneously,[16,17] the regulation of even a handful of miRNAs by the TGFβ/BMP signaling pathway could have a broad impact on gene expression and cellular physiology. As an increasing number of molecules and signaling pathways that affect miRNA biosynthesis are uncovered, a major challenge is to elucidate how multiple mechanisms of miRNA regulation cooperatively control both the variety and level of miRNA expression in a context-dependent manner. It will also be crucial to understand the physiological impact of multiple miRNAs and their target mRNAs whose expression are orchestrated by the TGFβ/BMP-Smad pathway and associate them with biological processes known to be governed by the TGFβ/BMP signaling pathways.

ACKNOWLEDGEMENT

We are grateful to members of the Lagna and Hata laboratory for stimulating discussions and A. Hilyard, MC Chan, H. Kang, C. Wu and P. Nguyen for generating data on the miRNA studies. This work was supported by National Institute of Health grants (HL082854 and HL093154) and American Heart Association (0940095N) to A.H.

REFERENCES

1. Moustakas A, Heldin CH. The regulation of TGFβ signal transduction. Development 2009; 136:3699-714.
2. ten Dijke P, Arthur HM. Extracellular control of TGFβ signalling in vascular development and disease. Nature Rev Mol Cell Biol 2007; 8:857-68.
3. Massague J, Blain SW, Lo RS. TGFβ signaling in growth control, cancer and heritable disorders. Cell 2000; 103:295-309.
4. Massague J, Gomis RR. The logic of TGFβ signaling. FEBS Lett 2006; 580:2811-20.
5. Massague J, Seoane J, Wotton D. Smad transcription factors. Genes Dev 2005; 19:2783-810.
6. Zhang Y, Musci T, Derynck R. The tumor suppressor Smad4/DPC4 as a central mediator of Smad function. Curr Biol 1997; 7:270-6.
7. Bardeesy N, Cheng KH, Berger JH et al. Smad4 is dispensable for normal pancreas development yet critical in progression and tumor biology of pancreas cancer. Genes Dev 2006; 20:3130-46.
8. Giehl K, Imamichi Y, Menke A. Smad4-independent TGF-β signaling in tumor cell migration. Cells Tissues Organs 2007; 185:123-30.
9. Rodriguez A, Griffiths-Jones S, Ashurst JL et al. Identification of mammalian microRNA host genes and transcription units. Genome Res 2004; 14:1902-10.
10. Cai X, Hagedorn CH, Cullen BR. Human microRNAs are processed from capped, polyadenylated transcripts that can also function as mRNAs. Rna 2004; 10:1957-66.
11. Lee Y, Kim M, Han J et al. microRNA genes are transcribed by RNA polymerase II. EMBO J 2004; 23:4051-60.
12. Grimson A, Farh KK, Johnston WK et al. microRNA targeting specificity in mammals: determinants beyond seed pairing. Mol Cell 2007; 27:91-105.
13. Lewis BP, Burge CB, Bartel DP. Conserved seed pairing, often flanked by adenosines, indicates that thousands of human genes are microRNA targets. Cell 2005; 120:15-20.
14. Lytle JR, Yario TA, Steitz JA. Target mRNAs are repressed as efficiently by microRNA-binding sites in the 5′ UTR as in the 3′ UTR. Proc Natl Acad Sci USA 2007; 104:9667-72.

15. Forman JJ, Legesse-Miller A, Coller HA. A search for conserved sequences in coding regions reveals that the let-7 microRNA targets Dicer within its coding sequence. Proc Natl Acad Sci USA 2008; 105:14879-84.
16. Selbach M, Schwanhausser B, Thierfelder N et al. Widespread changes in protein synthesis induced by microRNAs. Nature 2008; 455:58-63.
17. Baek D, Villen J, Shin C et al. The impact of microRNAs on protein output. Nature 2008; 455:64-71.
18. Huppi K, Volfovsky N, Runfola T et al. The identification of microRNAs in a genomically unstable region of human chromosome 8q24. Mol Cancer Res 2008; 6:212-21.
19. Huppi K, Volfovsky N, Mackiewicz M et al. microRNAs and genomic instability. Semin Cancer Biol 2007; 17:65-73.
20. Kumar MS, Erkeland SJ, Pester RE et al. Suppression of nonsmall cell lung tumor development by the let-7 microRNA family. Proc Natl Acad Sci USA 2008; 105:3903-8.
21. Lu J, Getz G, Miska EA et al. microRNA expression profiles classify human cancers. Nature 2005; 435:834-8.
22. Lee Y, Ahn C, Han J et al. The nuclear RNase III Drosha initiates microRNA processing. Nature 2003; 425:415-9.
23. Han J, Lee Y, Yeom KH et al. The Drosha-DGCR8 complex in primary microRNA processing. Genes Dev 2004; 18:3016-27.
24. Han J, Lee Y, Yeom KH et al. Molecular basis for the recognition of primary microRNAs by the Drosha-DGCR8 complex. Cell 2006; 125:887-901.
25. Morlando M, Ballarino M, Gromak N et al. Primary microRNA transcripts are processed cotranscriptionally. Nat Struct Mol Biol 2008; 15:902-9.
26. Gregory RI, Yan KP, Amuthan G et al. The Microprocessor complex mediates the genesis of microRNAs. Nature 2004; 432:235-40.
27. Davis BN, Hilyard AC, Lagna G et al. SMAD proteins control DROSHA-mediated microRNA maturation. Nature 2008; 454:56-61.
28. Yamagata K, Fujiyama S, Ito S et al. Maturation of microrna is hormonally regulated by a nuclear receptor. Mol Cell 2009; 36:340-7.
29. Fukuda T, Yamagata K, Fujiyama S et al. DEAD-box RNA helicase subunits of the Drosha complex are required for processing of rRNA and a subset of microRNAs. Nat Cell Biol 2007; 9:604-11.
30. Fuller-Pace FV. DExD/H box RNA helicases: multifunctional proteins with important roles in transcriptional regulation. Nucleic Acids Res 2006; 34:4206-15.
31. Warner DR, Bhattacherjee V, Yin X et al. Functional interaction between Smad, CREB binding protein and p68 RNA helicase. Biochem Biophys Res Commun 2004; 324:70-6.
32. Fuentealba LC, Eivers E, Ikeda A et al. Integrating patterning signals: Wnt/GSK3 regulates the duration of the BMP/Smad1 signal. Cell 2007; 131:980-93.
33. Kretzschmar M, Doody J, Massagué J. Opposing BMP and EGF signalling pathway converge on the TGFβ family mediator Smad1. Nature 1997; 389:618-22.
34. Lagna G, Hata A, Hemmati-Brivanlou A et al. Partnership between DPC4 and SMAD proteins in TGF-β signalling pathways. Nature 1996; 383:832-6.
35. Xu L, Yao X, Chen X et al. Msk is required for nuclear import of TGFβ/BMP-activated Smads. J Cell Biol 2007; 178:981-94.
36. Yao X, Chen X, Cottonham C et al. Preferential utilization of Imp7/8 in nuclear import of Smads. J Biol Chem 2008; 283:22867-74.
37. Qian B, Katsaros D, Lu L et al. High miR-21 expression in breast cancer associated with poor disease-free survival in early stage disease and high TGF-β1. Breast Cancer Res Treat 2009; 117:131-40.
38. Suzuki HI, Yamagata K, Sugimoto K et al. Modulation of microRNA processing by p53. Nature 2009; 460:529-33.
39. Sato MM, Nashimoto M, Katagiri T et al. Bone morphogenetic protein-2 down-regulates miR-206 expression by blocking its maturation process. Biochem Biophys Res Commun 2009; 383:125-9.
40. Miller SJ, Suthiphongchai T, Zambetti GP et al. p53 binds selectively to the 5' untranslated region of cdk4, an RNA element necessary and sufficient for transforming growth factor β- and p53-mediated translational inhibition of cdk4. Mol Cell Biol 2000; 20:8420-31.
41. Cassiday LA, Maher LJ 3rd. Having it both ways: transcription factors that bind DNA and RNA. Nucleic Acids Res 2002; 30:4118-26.
42. Fuller-Pace FV, Ali S. The DEAD box RNA helicases p68 (Ddx5) and p72 (Ddx17): novel transcriptional coregulators. Biochem Soc Trans 2008; 36:609-12.
43. Clark EL, Coulson A, Dalgliesh C et al. The RNA helicase p68 is a novel androgen receptor coactivator involved in splicing and is overexpressed in prostate cancer. Cancer Res 2008; 68:7938-46.
44. Endoh H, Maruyama K, Masuhiro Y et al. Purification and identification of p68 RNA helicase acting as a transcriptional coactivator specific for the activation function 1 of human estrogen receptor alpha. Mol Cell Biol 1999; 19:5363-72.

45. Yu B, Bi L, Zheng B et al. The FHA domain proteins DAWDLE in Arabidopsis and SNIP1 in humans act in small RNA biogenesis. Proc Natl Acad Sci USA 2008; 105:10073-8.
46. Ozsolak F, Poling LL, Wang Z et al. Chromatin structure analyses identify miRNA promoters. Genes Dev 2008; 22:3172-83.
47. Corcoran DL, Pandit KV, Gordon B et al. Features of mammalian microRNA promoters emerge from polymerase II chromatin immunoprecipitation data. PLoS One 2009; 4:e5279.
48. Kato M, Putta S, Wang M et al. TGF-β activates Akt kinase through a microRNA-dependent amplifying circuit targeting PTEN. Nat Cell Biol 2009; 11:881-9.
49. Qi HH, Ongusaha PP, Myllyharju J et al. Prolyl 4-hydroxylation regulates Argonaute 2 stability. Nature 2008; 455:421-4.
50. Diederichs S, Haber DA. Dual role for argonautes in microRNA processing and post-transcriptional regulation of microRNA expression. Cell 2007; 131:1097-108.
51. Chen L, Shen YH, Wang X et al. Human prolyl-4-hydroxylase alpha(I) transcription is mediated by upstream stimulatory factors. J Biol Chem 2006; 281:10849-55.

CHAPTER 3

STIMULATION OF pri-miR-18a PROCESSING BY hnRNP A1

Gracjan Michlewski, Sonia Guil and Javier F. Cáceres*

Abstract: Recent evidence suggests that the canonical miRNA processing pathway can be regulated by a number of positive and negative trans-acting factors. This chapter provides an overview of hnRNP A1-mediated regulation of miR-18a biogenesis. Our laboratory has recently established that the multifunctional RNA-binding protein hnRNP A1 is required for the processing of miR-18a at the nuclear step of Drosha-mediated processing. By combining structural and functional analysis of RNA, we showed that hnRNP A1 regulates the processing of pri-miR-18a by binding to its terminal loop and reshaping its stem-loop structure, thus allowing for a more effective Drosha cleavage. Furthermore, we linked the event of hnRNP A1-binding to the pri-miR-18a with an unusual phylogenetic sequence conservation of its terminal loop. Bioinformatic and mutational analysis revealed that a number of pri-miRNAs have highly conserved terminal loops, which are predicted to act as landing pads for trans-acting factors influencing miRNA processing. These results underscore a previously uncharacterized role for general RNA-binding proteins as factors that facilitate the processing of specific miRNAs, revealing an additional level of complexity for the regulation of miRNA production and function.

INTRODUCTION

Mature miRNAs are derived from primary transcripts (pri-miRNAs) by sequential nuclear and cytoplasmic processing events (reviewed by ref. 1). In the nucleus, the microprocessor complex comprising the RNase III-Type enzyme Drosha and its RNA-binding partner DGCR8 generates stem-loop precursors, termed pre-miRNAs,[2-6] which are exported to the cytoplasm by Exportin 5.[7-9] Subsequently, there is an additional cytoplasmic processing event carried out by the Type III ribonuclease Dicer, which results

*Corresponding Author: Javier F. Cáceres—Medical Research Council Human Genetics Unit, Institute of Genetics and Molecular Medicine,Western General Hospital, Edinburgh EH4 2XU, UK.
Email: javier.cáceres@hgu.mrc.ac.uk

Regulation of microRNAs, edited by Helge Großhans.
©2010 Landes Bioscience and Springer Science+Business Media.

in the production of mature miRNAs.[10-13] One strand of this duplex is then incorporated into the RNA-Induced Silencing Complex (RISC), which targets specific mRNAs and controls their expression by either affecting transcript stability or translation (reviewed by refs. 14,15). Although the canonical pathway of microRNA biogenesis is well understood, diverse regulatory mechanisms governing post-transcriptional processing of microRNAs are beginning to emerge. This regulation can be accomplished at different levels in this pathway affecting Drosha or Dicer-mediated processing.[16] Furthermore, only recently it was established that RNA turnover acts to modulate mature miRNA levels and activity in the nematode *C.elegans* (see chapter by Großhans and Chatterjee).[17] The DEAD-box helicases p68/p72, which are components of the large microprocessor complex, have been found to be required for the processing of a subset of miRNAs.[18] They also seem to act to recruit regulators of Drosha-mediated processing, such as the estrogen receptor[19,20] (see also chapter by Fujiyama-Nakamura et al) and components of the transforming growth factor beta (TGF-β) signaling (see also chapter by Hata and Davis).[21]

hnRNP A1 BINDS TO THE pri-miR-18a STEM-LOOP STRUCTURE

A family of RNA-binding proteins, collectively known as heterogenous nuclear ribonucleoproteins (hnRNP proteins), have a role in almost every aspect of mRNA metabolism, including transcription and pre-mRNA splicing in the nucleus, mRNA export and also many cytoplasmic events.[22,23] Among them, the nucleo-cytoplasmic shuttling protein hnRNP A1 has been extensively studied. In the nucleus, hnRNP A1 and other members of the hnRNP A/B family of proteins has been shown to affect alternative splicing regulation by antagonizing the function of the SR family of proteins[24-27] and also to have a function in constitutive splicing by modulating the conformation of mammalian pre-mRNAs.[28] In the cytoplasm, hnRNP A1 regulates internal ribosome entry site (IRES)-mediated translation.[29-31] Surprisingly, only a relatively small number of cellular and viral genes regulated by hnRNP A1 have been identified, which prompted us to search for endogenous RNA targets for this protein in HeLa cells. We used a Cross-Linking Immunoprecipitation protocol (CLIP),[32,33] which relies on an in vivo UV cross-linking step that is followed by highly stringent immunoprecipitation conditions, so that only endogenous RNAs directly bound to the protein of interest are selected. This resulted in the identification of a few hundred mRNA targets and also a microRNA precursor, pri-miR-18a,[34] which is expressed as part of the miR-17-92 cluster of intronic miRNAs. This cluster contains six precursor miRNAs on chromosome 13 (mir-17, mir-18a, mir-19a, mir-20a, mir-19b-1 and mir-92)[35] but hnRNP A1 was found to bind only to miR-18a.[34] This cluster has been shown to be amplified in human B-cell lymphomas[36] and its overexpression promotes c-myc-induced tumor development in a mouse B-cell lymphoma model.[37] The CLIP experiment revealed that the sequence bound by hnRNP A1 corresponded to the stem-loop precursor miR-18a, which after processing renders two mature miRNAs: miR-18a and miR-18a*. This binding was confirmed using Immunoprecipitation and RT-PCR (IP-RT-PCR) and RNA chromatography and revealed that hnRNP A1 binds pri-miR-18a prior to Drosha cleavage.[34]

hnRNP A1 PROMOTES THE DROSHA-MEDIATED PROCESSING
OF pri-miR-18a

The binding of hnRNP A1 to pri-miR-18a was found to correlate with increased levels of mature miR-18a, indicating a putative role of hnRNP A1 in the biogenesis of this miRNA. Several lines of evidence confirmed this. First, the levels of pre-miR-18a were strongly reduced in HeLa cells depleted of hnRNP A1, as revealed in Northern blot analysis. Secondly, hnRNP A1 was shown to be required for miR-18a-mediated repression of a target reporter in vivo. Thirdly, HeLa cell extracts depleted of hnRNP A1 have reduced in vitro processing activity of pri-miRNA-18a and also display reduced levels of endogenous pre-miR-18a.[34] Importantly, depletion of hnRNP A1 had no impact on the efficient processing of a control miRNA cluster, nor did it affect the processing of other members of the miR-17-92 cluster, indicating that hnRNP A1 specifically binds to, and affects the processing of, miR-18a. Interestingly, placing pri-miR-18a in a different context, makes its processing independent of hnRNP A1, suggesting that the sequence and natural context of pri-miR-18a constitutes a suboptimal recognition site for Drosha/ DGCR8 cropping. Altogether, this strongly suggested a direct role for hnRNP A1 in the biogenesis of miR-18a.[34]

RNA footprint analysis using pri-miR-18a and recombinant hnRNP A1 protein revealed two major binding sites of hnRNP A1: one corresponding to the terminal loop of pri-miR-18a and a secondary site that corresponded to the lower part of the stem[38] (Fig. 1). Binding to the latter resulted in the relaxation of the pri-miR-18a secondary structure, creating a more favorable cleavage site for Drosha (Fig. 1B). This probably reflects the reported unwinding/annealing activities of hnRNP A1.[39-41] Interestingly, both sites share similarity with the consensus hnRNP A1 binding site, UAGGGA/U as identified in SELEX experiments by the Dreyfuss laboratory.[42]

There is a highly related pri-miRNA sequence, pri-miR-18b, which is part of the homologous primary cluster miR106a~18b~20b located on chromosome X[35] and was shown to be processed independently of hnRNP A1.[34] Therefore, a comparison of the secondary structure and requirement for hnRNP A1 between pri-miR-18a and pri-miR-18b presented an ideal situation to verify the model displayed in Figures 1 and 2. Interestingly, pri-miR-18b in the absence of hnRNP A1 mimics the structural rearrangements seen in the stem of pri-miR-18a in the presence of added recombinant hnRNP A1 protein, which is the relaxation of residues between U56 and U60 that are involved in strong Watson-Crick pairing. Furthermore, forcing structural changes in the stem of pri-miR-18a, such that it presents a bulge in its stem, as naturally found in pri-miR-18b, made its processing more efficient and completely independent of the presence of hnRNP A1. Binding of hnRNP A1 to the terminal loop of pri-miR-18a was confirmed in EMSA analysis. Furthermore, mutations in the terminal loop of pri-miR-18a that eliminate any hnRNP A1-binding site blocked its in vitro processing in HeLa cell extracts. We speculate that hnRNP A1 binding to the terminal loop is required for efficient binding of hnRNP A1 at the base of the stem, which creates a more favorable Drosha-mediated processing site. Altogether, these experiments strongly suggest that the function of hnRNP A1 in the Drosha-mediated processing of pri-miR-18a is to bind and alter the local conformation of the stem in the vicinity of Drosha cleavage sites (Fig. 2).

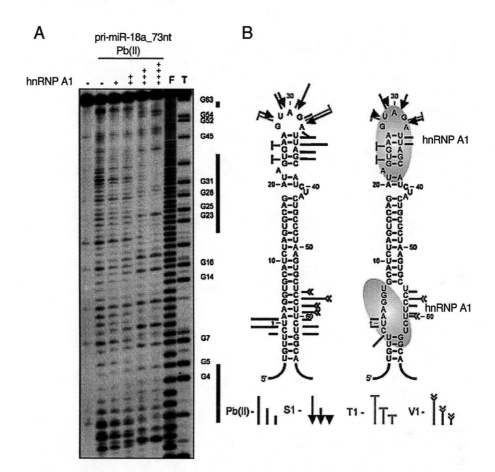

Figure 1. hnRNP A1 binds to the terminal and internal loops of pri-miR-18a causing relaxation of the stem. A) Footprint analysis of the pri-miR-18a/hnRNP A1 complex. Cleavage patterns were obtained for 5'[32]P-labeled pri-miR-18a transcript (100 × 10[3] c.p.m.) incubated in the presence of increasing concentrations of recombinant hnRNP A1 protein, as previously described.[36] F and T identify nucleotide residues subjected to partial digest with formamide (every nucleotide) or ribonuclease T1 (G-specific cleavage), respectively. Thick lines on the right-hand side indicate rows of nucleotides protected by hnRNP A1. Positions of selected residues are indicated (B) Proposed structure of free and hnRNP A1-bound pri-miR-18a. The sites and intensities of cleavages generated by structure probes located at the places of hnRNP A1 binding are shown. Nucleotides are numbered from the 5' site of Drosha cleavage. [Reproduced from: Michlewski G, Guil S, Semple C, Cáceres JF. Post-transcriptional regulation of miRNAs harboring conserved terminal loops. Mol Cell 2008; 32:383-393; ©2008 with permission from Elsevier.]

ROLE OF THE TERMINAL LOOPS IN miRNA PROCESSING

Binding of hnRNP A1 to the terminal loop of pri-miR-18a prompted us to revise the prevailing notion that the terminal loops of pri-miRNAs were not required for miRNA processing.[43] Interestingly, we found that the terminal loop of pri-miR-18a was found to be very well conserved across vertebrate species, together with other 73 miRNAs, representing 14% of all miRNAs analyzed (Fig. 3). This conservation was proposed to be diagnostic of these terminal loops acting as platform for the binding of auxiliary factors

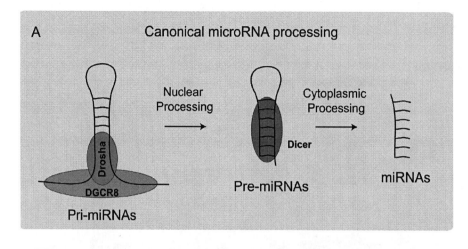

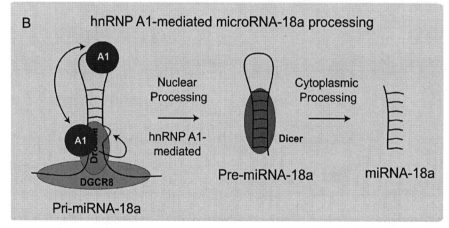

Figure 2. Model displaying the mechanism by which hnRNP A1 facilitates the Drosha-mediated processing of pri-miR-18a. A) Canonical pri-miRNA processing pathway. B) HnRNP A1-driven regulation of pri-miR-18a Drosha cleavage.

required for an efficient processing of this subset of pri-miRNAs. This was tested with the use of oligonucleotides complementary to conserved terminal loops, which were termed LooptomiRs (for Loop Targeting Oligonucleotide anti-miRNAs). It was shown that specific LooptomiRs block pri-miRNA processing of miRNAs with conserved terminal loops, whereas they had no effect on selected nonconserved terminal loops.[38] These findings suggest a general role of RNA-binding proteins as auxiliary factors in the microRNA processing pathway. Of course, binding of proteins to conserved terminal loops can result in an enhancement of miRNA biogenesis, as was demonstrated in the case of miR-18a, but could alternatively result in negative regulation of miRNA biogenesis, acting at the level of either Drosha or Dicer-mediated processing events, as shown in the case of Lin28-mediated negative regulation of let-7a biogenesis (see also chapter by Lehrbach and Miska).[16]

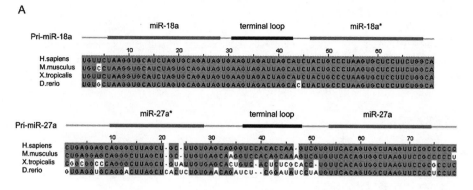

B

Pri-miRs with highly conserved terminal loops

miR-452	miR-323	miR-376a-1	miR-193b	let-7a-2
miR-106b	miR-34c	miR-375	miR-190	let-7a-1
miR-425	miR-96	miR-365-1	miR-181b-1	miR-30c-2
miR-133a-1	miR-9-2	miR-331	miR-153-2	miR-628
miR-181a-1	miR-9-1	miR-32	miR-142	miR-18b
miR-125b-1	miR-876	miR-31	miR-138-2	miR-10a
miR-582	miR-873	miR-29c	miR-135a-2	miR-30a
miR-18a	miR-7-1	miR-29b-2	miR-128a	miR-874
miR-103-1	miR-592	miR-29b-1	miR-107	miR-147b
miR-148a	miR-423	miR-29a	miR-101-1	let-7c
miR-19a	miR-411	miR-25	let-7i	miR-136
miR-1-2	miR-383	miR-22	let-7g	miR-361
miR-30c-1	miR-379	miR-21	let-7f-2	miR-101-2
miR-140	miR-378	miR-204	let-7f-1	miR-15a
miR-382	miR-377	miR-196b	let-7d	

Figure 3. A) The terminal loop of pri-miR-18a is atypically well conserved across vertebrate species as revealed by sequence alignments across four species for pri-miR-18a (upper panel) or pri-miR-27a (lower panel). The alignments were performed using ClustalW2 and visualised with BLOSUM62 Score. B) Phylogenetic analysis of human pri-miRNAs sequences shows that ~14% (74 out of 533) of the miRNAs analyzed had similar high conservation pattern across the whole pri-miRNA sequence, including their terminal loops.

CONCLUSION AND FUTURE PROSPECTS

In summary, experiments described in this chapter describe the function of a general RNA-binding protein, hnRNP A1, as a positive regulator of miR-18a processing (and perhaps of other miRNAs), acting at the level of Drosha-mediated processing. Furthermore, it emphasizes the role of conserved terminal-loops of pri-miRNAs as sites of regulation by trans-acting factors that could positively or negatively regulate miRNA biogenesis. Future directions will aim to identify additional regulatory factors binding to conserved terminal loops and establish their role in miRNA biogenesis pathway. It will be important to link the patterns of expression of these regulatory factors with the production of mature miRNAs and their effect on mRNA targets. Altogether, these findings open up new avenues towards a deeper understanding of miRNA biogenesis pathways and their contribution to physiological states and pathological conditions.

ACKNOWLEDGEMENTS

This work was supported by the MRC, a project grant from the Wellcome Trust and Eurasnet (European Alternative splicing Network-FP6). S.G is the recipient of a Ramon y Cajal contract from the MICINN (Spanish Government).

REFERENCES

1. Kim VN, Han J, Siomi MC. Biogenesis of small RNAs in animals. Nat Rev Mol Cell Biol 2009; 10:126-139.
2. Denli AM, Tops BB, Plasterk RH et al. Processing of primary microRNAs by the Microprocessor complex. Nature 2004; 432:231-235.
3. Gregory RI, Yan KP, Amuthan G et al. The Microprocessor complex mediates the genesis of microRNAs. Nature 2004; 432:235-240.
4. Han J, Lee Y, Yeom KH et al. The Drosha-DGCR8 complex in primary microRNA processing. Genes Dev 2004; 18:3016-3027.
5. Landthaler M, Yalcin A, Tuschl T. The human DiGeorge syndrome critical region gene 8 and Its D. melanogaster homolog are required for miRNA biogenesis. Curr Biol 2004; 14:2162-2167.
6. Zeng Y, Yi R, Cullen BR. Recognition and cleavage of primary microRNA precursors by the nuclear processing enzyme Drosha. EMBO J 2005; 24:138-148.
7. Yi R, Qin Y, Macara IG et al. Exportin-5 mediates the nuclear export of pre-microRNAs and short hairpin RNAs. Genes De 2003; 17:3011-3016.
8. Bohnsack MT, Czaplinski K, Gorlich D. Exportin 5 is a RanGTP-dependent dsRNA-binding protein that mediates nuclear export of pre-miRNAs. RNA 2004; 10:185-191.
9. Lund E, Guttinger S, Calado A et al. Nuclear export of microRNA precursors. Science 2004; 303:95-98.
10. Grishok A, Pasquinelli AE, Conte D et al. Genes and mechanisms related to RNA interference regulate expression of the small temporal RNAs that control C. elegans developmental timing. Cell 2001; 106:23-34.
11. Hutvagner G, McLachlan J, Pasquinelli AE et al. A cellular function for the RNA-interference enzyme Dicer in the maturation of the let-7 small temporal RNA. Science 2001; 293:834-838.
12. Ketting RF, Fischer SE, Bernstein E et al. Dicer functions in RNA interference and in synthesis of small RNA involved in developmental timing in C. elegans. Genes Dev 2001; 15:2654-2659.
13. Knight SW, Bass BL. A role for the RNase III enzyme DCR-1 in RNA interference and germ line development in Caenorhabditis elegans. Science 2001; 293:2269-2271.
14. Bartel DP. microRNAs: target recognition and regulatory functions. Cell 2009; 136:215-233.
15. Filipowicz W, Bhattacharyya SN, Sonenberg N. Mechanisms of post-transcriptional regulation by microRNAs: are the answers in sight? Nat Rev Genet 2008; 9:102-114.
16. Winter J, Jung S, Keller S et al. Many roads to maturity: microRNA biogenesis pathways and their regulation. Nat Cell Biol 2009; 11:228-234.
17. Chatterjee S, Grosshans H. Active turnover modulates mature microRNA activity in Caenorhabditis elegans. Nature 2009; 461:546-549.
18. Fukuda T, Yamagata K, Fujiyama S et al. DEAD-box RNA helicase subunits of the Drosha complex are required for processing of rRNA and a subset of microRNAs. Nat Cell Biol 2007; 9:604-611.
19. Macias S, Michlewski G, Caceres JF. Hormonal regulation of microRNA biogenesis. Mol Cell 2009; 36:172-173.
20. Yamagata K, Fujiyama S, Ito S et al. Maturation of microRNA Is Hormonally Regulated by a Nuclear Receptor. Mol Cell 2009; 36:340-347.
21. Davis BN, Hilyard AC, Lagna G et al. SMAD proteins control DROSHA-mediated microRNA maturation. Nature 2008; 454:56-61.
22. Dreyfuss G, Kim VN, Kataoka N. Messenger-RNA-binding proteins and the messages they carry. Nat Rev Mol Cell Biol 2002; 3:195-205.
23. Martinez-Contreras R, Cloutier P, Shkreta L et al. hnRNP proteins and splicing control. Adv Exp Med Biol 2007; 623:123-147.
24. Mayeda A, Krainer AR. Regulation of alternative pre-mRNA splicing by hnRNP A1 and splicing factor SF2. Cell 1992; 68:365-375.
25. Mayeda A, Munroe SH, Caceres JF et al. Function of conserved domains of hnRNP A1 and other hnRNP A/B proteins. EMBO J 1994; 13:5483-5495.

26. Caceres JF, Stamm S, Helfman DM et al. Regulation of alternative splicing in vivo by overexpression of antagonistic splicing factors. Science 1994; 265: 1706-1709.
27. Yang X, Bani MR, Lu SJ et al. The A1 and A1B proteins of heterogeneous nuclear ribonucleoparticles modulate 5' splice site selection in vivo. Proc Natl Acad Sci USA 1994; 91:6924-6928.
28. Martinez-Contreras R, Fisette JF, Nasim FU et al. Intronic Binding Sites for hnRNP A/B and hnRNP F/H Proteins Stimulate Pre-mRNA Splicing. PLoS Biol 2006; 4:e21.
29. Bonnal S, Pileur F, Orsini C et al. Heterogeneous nuclear ribonucleoprotein A1 is a novel internal ribosome entry site trans-acting factor that modulates alternative initiation of translation of the fibroblast growth factor 2 mRNA. J Biol Chem 2005; 280:4144-4153.
30. Cammas A, Pileur F, Bonnal S et al. Cytoplasmic relocalization of heterogeneous nuclear ribonucleoprotein A1 controls translation initiation of specific mRNAs. Mol Biol Cell 2007; 18:5048-5059.
31. Jo OD, Martin J, Bernath A et al. Heterogeneous nuclear ribonucleoprotein A1 regulates cyclin D1 and c-myc internal ribosome entry site function through Akt signaling. J Biol Chem 2008; 283:23274-23287.
32. Ule J, Jensen KB, Ruggiu M et al. CLIP identifies Nova-regulated RNA networks in the brain. Science 2003; 302:1212-1215.
33. Ule J, Jensen K, Mele A et al. CLIP: A method for identifying protein-RNA interaction sites in living cells. Methods 2005; 37:376-386.
34. Guil S, Caceres JF. The multifunctional RNA-binding protein hnRNP A1 is required for processing of miR-18a. Nat Struct Mol Biol 2007; 14:591-596.
35. Tanzer A, Stadler PF. Molecular evolution of a microRNA cluster. J Mol Biol 2004; 339:327-335.
36. Ota A, Tagawa H, Karnan S et al. Identification and characterization of a novel gene, C13orf25, as a target for 13q31-q32 amplification in malignant lymphoma. Cancer Res 2004; 64:3087-3095.
37. He L, Thomson JM, Hemann MT et al. A microRNA polycistron as a potential human oncogene. Nature 2005; 435:828-833.
38. Michlewski G, Guil S, Semple CA et al. Post-transcriptional regulation of miRNAs harboring conserved terminal loops. Mol Cell 2008; 32:383-393.
39. Kumar A, Wilson SH. Studies of the strand-annealing activity of mammalian hnRNP complex protein A1. Biochemistry 1990; 29:10717-10722.
40. Pontius BW, Berg P. Renaturation of complementary DNA strands mediated by purified mammalian heterogeneous nuclear ribonucleoprotein A1 protein: implications for a mechanism for rapid molecular assembly. Proc Natl Acad Sci USA 1990; 87:8403-8407.
41. Munroe SH, Dong XF. Heterogeneous nuclear ribonucleoprotein A1 catalyzes RNA.RNA annealing. Proc Natl Acad Sci USA 1992; 89:895-899.
42. Burd CG, Dreyfuss G. RNA binding specificity of hnRNP A1: significance of hnRNP A1 high-affinity binding sites in pre-mRNA splicing. EMBO J 1994; 13:1197-1204.
43. Han J, Lee Y, Yeom KH et al. Molecular Basis for the Recognition of Primary microRNAs by the Drosha-DGCR8 Complex. Cell 2006; 125: 887-901.

CHAPTER 4

KSRP PROMOTES THE MATURATION
OF A GROUP OF miRNA PRECURSORS

Michele Trabucchi,* Paola Briata, Witold Filipowicz, Andres Ramos,
Roberto Gherzi and Michael G. Rosenfeld

Abstract: microRNAs (miRNAs) are small noncoding RNAs that down-regulate gene expression
by reducing stability and/or translation of target mRNAs. In animals, miRNAs
arise from sequential processing of hairpin primary transcripts by two RNAse III
domain-containing enzymes, namely Drosha and Dicer, to generate a mature form
of about 22 nucleotides. In this chapter we discuss our latest findings indicating
that KSRP is an integral component of both Drosha and Dicer complexes. KSRP
binds to the terminal loop sequence of a subset of miRNA precursors promoting
their maturation. Our data indicate that the terminal loop is a pivotal structure where
activators of miRNA processing as well as repressors of miRNA processing act in
a coordinated way to convert cellular signals into changes in miRNA expression
processing. This uncovers a new level of complexity of miRNA mechanisms for
gene expression regulation.

INTRODUCTION

miRNA precursors are encoded within diverse functional regions of the genome
with more than half of human miRNA genes encoded within introns, while others
are transcribed as independent monocistronic or polycistronic units.[1] The majority of
miRNAs are transcribed by RNA polymerase II and, subsequently, they are capped and
polyadenylated.[1,2] It has been recently demonstrated that Drosha cleavage of intronic
pri-miRNAs occurs cotranscriptionally and precedes splicing.[3] Additional work from

*Corresponding Author: Michele Trabucchi—Howard Hughes Medical Institute, Department and School
of Medicine, University of California, San Diego, 9500 Gilman Drive, Room 345, La Jolla, California
92093-0648, USA. Email: mtrabucc@ucsd.edu

Regulation of microRNAs, edited by Helge Großhans.
©2010 Landes Bioscience and Springer Science+Business Media.

Bozzoni and coworkers further strengthened the connection between transcription and Drosha cleavage also for intragenic pri-miRNAs.[4]

The evolutionarily-conserved mechanism by which primary miRNAs (pri-miRNAs) are processed first to precursor miRNAs (pre-miRNAs) and then to mature miRNAs involves two consecutive endonucleolytic cleavages executed by multiprotein complexes containing the RNase III enzymes Drosha and Dicer, respectively.[5] Drosha processes the pri-miRNA into a ~70 nt hairpin pre-miRNA.[5] Through the interaction with Exportin-5 and Ran-GTP, the pre-miRNA is transported into the cytoplasm, where it undergoes a second round of processing catalyzed by Dicer.[5] This cleavage event gives rise to a double-stranded ~22 nt product composed of the mature miRNA guide strand and the miRNA* (star) passenger strand.[5] The mature miRNA is then loaded into the RNA-Induced Silencing Complex (RISC) while the passenger strand is degraded. In the context of the RISC, miRNAs posttranscriptionally regulate the expression of target genes.[5,6] Although major progress has been made in understanding the basic mechanism of miRNA biogenesis, many questions remain unanswered. Specific chapters of this book, besides recently published reviews, extensively describe the majority of the known aspects of miRNA biogenesis regulation[1,2,7] Thus we will focus our discussion on some open questions regarding mechanistic and functional aspects of miRNA biogenesis controlled by the multifunctional single-strand RNA-binding protein KSRP (KH-type splicing regulatory protein).

CO-ACTIVATORS AND CO-REPRESSORS OF miRNA PRECURSOR MATURATION

Pioneering studies from different laboratories have envisaged the possibility that miRNA maturation is a finely regulated process that involves co-regulators and is suited to respond to changing cellular conditions.[8-13]

Recent studies have revealed that different regulatory proteins participate in the specific control of individual miRNAs. The DEAD-box RNA helicases p68 (DDX5) and p72 (DDX17) were identified as components of the Drosha-mediated processing complex and, in the absence of both DDX5 and DDX17 the expression of approximately 35% of miRNAs (and pre-miRNAs) was reduced without concomitant changes in the levels of corresponding pri-miRNAs.[9] This suggested a role for DDX5/DDX17 in promoting the Drosha-mediated processing of a subset of miRNAs. The precise mechanism of DDX5/DDX17-mediated processing is still unclear, but may involve re-arrangement of the RNA hairpin, which results in enhanced Drosha recruitment or stability. Alternatively, as DDX5/DDX17 are known to interact with a variety of proteins, they may serve as a scaffold for the recruitment of multiple factors to the Drosha complex. Indeed, a recent paper revealed that the tumor suppressor and transcription factor p53 interacts with the Drosha complex through the association with DDX5 and facilitates the maturation of a restricted population of pri-miRNAs in response to DNA damage in cancer cells.[14] The role of DDX5/DDX17 as adaptors for signal- and cell- specific cofactors in miRNA processing is further supported by the positive regulation of Drosha-mediated processing mediated by the DDX5-interacting Smad proteins, the signal transducers of the TGF-beta family signaling cascade (see also chapter by Hata and Davis).[8] Very recent data from the Kato laboratory, reviewed in the chapter by Fujiyama-Nakamura et al, suggested a mechanism whereby estradiol-bound estrogen receptor α blocks Drosha-mediated processing of a subset of miRNAs by binding to Drosha in a DDX5/DDX17-dependent manner and inducing the dissociation of the microprocessor complex from the pri-miRNA.[15]

Smad proteins as well as p53 and estrogen receptor α do not directly bind to RNA. Conversely, other Drosha cofactors are RNA-binding proteins. The heterogeneous nuclear ribonucleoprotein (hnRNP) A1, which interacts with RNA through its RNA recognition motifs and appears to be involved in many aspects of RNA life, originally proved to be required for the maturation of miR-18a, a member of the miR-17-92 cluster.[10] We recently demonstrated that also the KH-type splicing regulatory protein (KSRP) is able to interact with select miRNA precursors.[12] KSRP is a multifunctional single–strand RNA-binding protein that affects many steps of RNA life including splicing, localization and degradation.[16-20] Recently, we demonstrated that KSRP binds to the terminal loop (TL) of a cohort of miRNA precursors and interacts with both Drosha and Dicer to promote miRNA maturation.[12,21] In Figure 1, the miRNAs whose biogenesis is regulated by KSRP are listed. Interestingly, also hnRNP A1 binds to the TL of a group of pri-miRNAs, which partly overlaps that interacting with KSRP.[10,22] Even more intriguingly, KSRP interacts with hnRNP A1 and regulates its expression by affecting its mRNA half-life.[23] A relevant difference between hnRNP A1- and KSRP-activated miRNA maturation is that, in contrast to hnRNP A1, KSRP functions not only in the nuclear maturation of pri-miRNAs to pre-miRNAs but also in the cytoplasmic maturation of pre-miRNAs into miRNAs, thus representing a link between nuclear and cytoplasmic events. Indeed, we also obtained evidence that KSRP interacts with Exportin-5 in an RNase A-sensitive way thus suggesting that KSRP is associated with the TL of target miRNA precursors during nucleo-cytoplasmic transit.

KSRP-RNA recognition is a complex event and relies both on the sequence selectivity of the KH domains and on the actual availability of single stranded RNA

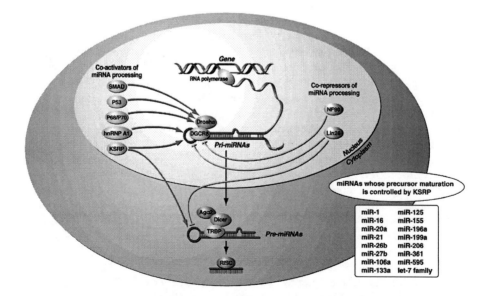

Figure 1. A schematic model of the interplay between positive and negative co-regulators of miRNA precursors maturation, including a list of miRNAs whose maturation is controlled by KSRP.

sequences at the recognition site.[17,24] We have examined the sequence preference of the four KH domains of KSRP. Three of them (KH1, KH2 and KH4) display a moderate selectivity towards specific sequences, while KH3 can discriminate strongly in favour of short G-rich stretches.[24] If a G-rich stretch is available within the single stranded RNA target (e.g., in the let-7a precursors) KH3 will bind to it, defining KSRP's binding frame. However, if the RNA target does not include a G-rich stretch, the four KH domains of KSRP will explore all available sequences in order to optimize the binding affinity of KSRP (e.g., in the miR-21 precursors). In either case, the structural setting of the single-stranded sequences will play a major role in KSRP binding, as KSRP-RNA interaction is a multi-domain event and steric hindrance may limit the space available to the protein domains within the RNA 3D structure. The size and conformation of the TL of miRNA precursors vary significantly and, together with the presence of a G-rich sequence, could represent a powerful selector for KSRP recognition.[12] In conclusion, we propose a general model for KSRP–RNA interactions based on the differential use of multiple domains that underscores the adaptability of the protein to a broad range of single-strand RNA sequences.

A number of manuscripts appeared in the last year proving that miRNA maturation can also be negatively regulated through the intervention of cofactors. We have already mentioned the estrogen receptor α, which exerts an indirect negative function on miRNA maturation acting through DDX5/DDX17.[15] The RNA-binding protein Lin-28 is able to repress the maturation of let-7 family members and this effect is mediated by its interaction with the TL of let-7 precursors (see also chapter by Lehrbach and Miska).[13,25,26] In addition to inhibition of the Drosha-mediated processing step, Lin-28 also inhibits the Dicer-mediated processing of let-7 family members.[13,27,28] Indeed, similarly to KSRP, Lin-28 is a shuttling protein and it is abundant in the cytoplasm, suggesting that this may be the primary location of its interaction with let-7 precursors. Data from Wulczyn laboratory suggest that Lin-28 compete with Dicer for access to pre-let-7.[27] Additionally, Kim and coworkers recently reported that Lin-28 promotes the 3'-uridylation of pre-let-7, which inhibits Dicer-mediated processing and leads to degradation of pre-let-7 itself.[29-31]

While Lin-28 repressor function seems to be restricted to let-7 family members, the double-stranded RNA-binding proteins NF90 and NF45 reduce the production of a broader spectrum of pre- and mature miRNAs. NF90 and NF45 interact with the stem of a group of pri-miRNAs to preclude their binding to DGCR8, an essential member of the Drosha complex and, in turn, their maturation into pre-miRNAs.[32]

Figure 1 summarizes the proposed scenario of activators and repressors of miRNA maturation.

IMPACT OF KSRP AND OTHER CO-ACTIVATORS AND CO-REPRESSORS OF miRNA PRECURSOR MATURATION ON CELL PROLIFERATION, DIFFERENTIATION AND CANCER

As a single miRNA modulates the expression of many targets simultaneously, it is able to rapidly modify complex cellular functions that require the coordinated regulation of gene networks, in response to environmental cues. Therefore, regulation of miRNA biogenesis may serve as an important line of response to promote the modulation of gene expression programs in response to cellular stimuli. As KSRP as well as the majority of co-regulators mentioned in this review are able to influence maturation of discrete groups of miRNAs, this could allow the coregulation of miRNAs implicated in certain cellular functions in response to certain cellular stimuli.

A good example of how the control of maturation of a group of miRNAs by a single cofactor can affect a cellular function is represented by the KSRP-directed maturation of myogenic miRNAs in response to differentiative stimuli in myoblasts.[12] Upon serum withdrawal, C2C12 myoblasts undergo differentiation into myotubes. This event is accompanied by enhanced expression of some "myogenic" miRNAs (including miR-1, miR-133a and miR-206) reported to have a causative role in the differentiation process.[33,34] KSRP knockdown impairs myogenic miRNA maturation, increases the expression of some of their targets and inhibits C2C12 myoblasts differentiation.[12] These data link the stimulation of pri-miRNA processing by KSRP to a mammalian differentiation program.

Some evidence suggests a functional interplay between positive and negative co-regulators of pri-miRNA maturation. Lin-28 and mature let-7g show reciprocal expression patterns during both embryonic development and embryonic stem cell differentiation thus supporting a role of Lin-28 in let-7g regulation during embryogenesis.[1,2] Even though Lin-28 and KSRP do not share common binding sites in the TL of let-7 family precursors,[12,25,26] our data suggest that when Lin-28 is expressed in undifferentiated embryonic carcinoma cells, KSRP cannot interact with pri-let-7g. When Lin28 is not expressed, as in differentiated P19 cells or in undifferentiated P19 upon specific Lin-28 knockdown, KSRP is able to promote let-7g maturation.[12] We propose that the TL is a pivotal structure where miRNA-processing-co-activators (KSRP and possibly additional RNA-binding proteins) as well as miRNA-processing co-repressors (as exemplified by Lin-28 for let-7 and possibly additional RNA-binding proteins) function in a coordinated way to convert proliferation and differentiation cues into changes of miRNA expression. In other words, the occurrence of a co-activator and a co-repressor for regulation of miRNA maturation extends the concept of opposing co-regulators, analogous to events now well established for DNA-binding transcription factors (Fig. 1).

The differential expression of miRNA processing co-regulators, as well as their posttranslational modifications and sub-cellular localization, may serve to regulate miRNA expression in a tissue-or context-dependent manner. For example, phosphorylation of hnRNP A1 by MAPK p38 has been reported to promote cytoplasmic localization of hnRNP A1.[35,36] Similarly, KSRP can be phosphorylated by two distinct Ser/Thr kinase, MAPK p38 and Akt. While MAPK p38-mediated phosphorylation affects the binding efficiency KSRP of to some RNA substrates, Akt promotes nuclear accumulation of KSRP through interaction with 14-3-3.[37-39] Intriguingly, phosphorylation by Akt impairs the ability of KSRP to interact with some enzymes responsible for decay of labile mRNAs.[39] An important future challenge will be to systematically dissect pathways that modulate the function of KSRP and other RNA-binding proteins in the regulation of miRNA biogenesis.

Furthermore, it is now clear that the function of a single cofactor can vary depending on the cellular context in which both the cofactor and its target miRNAs are expressed. For example KSRP, which is ubiquitously expressed, affects, through miRNA maturation control, cellular functions as different as myoblast differentiation and inflammatory response to microbic products (such as lipopolysaccharide, LPS) in macrophages. In this last case, the regulation of a single miRNA (miR-155) is responsible for the KSRP-mediated response to LPS.[21]

Finally, our studies showed that KSRP knockdown limits, in a let-7a-dependent way, cell proliferation by influencing the expression of let-7a targets such as MYC and NRAS.[12] A seminal study from Thomson and coworkers implicated the regulation of miRNA precursor processing in cell transformation and cancer.[40] Indeed, a notable global reduction of mature miRNAs has been observed in cancers.[41] In addition, the importance of miRNA processing regulation for tumorigenesis has been experimentally proved by Drosha, Dgcr8 or Dicer knockdown.[42] The demonstration that Lin-28 is overexpressed in primary human tumors and human cancer cell lines,[43-45] which is linked to the down-regulation of let-7 expression,[13] supports the idea that altered control of miRNA biogenesis may critically impact on cancer pathogenesis, representing a stimulus for further intense investigations.

CONCLUSION

Altogether, we conclude that KSRP also serves as a previously unsuspected component of both Drosha and Dicer complexes and regulates the biogenesis of a subset of miRNAs. KSRP binds in a sequence-specific fashion to the TL of a subset of pri- and pre-miRNAs and functions as a co-activator for miRNA processing. The evidence that both co-activators (such as KSRP) and co-repressors (such as Lin-28) of miRNA maturation exist and their interplay is required to precisely regulate miRNA expression in specific cellular contexts provides the rational basis to identify additional co-regulators of miRNA processing, stimulating therefore future research in this area of gene expression regulation.

REFERENCES

1. Davis BN, Hata A. Regulation of microRNA Biogenesis: A miRiad of mechanisms. Cell Commun Signal 2009; 7(1):18.
2. Winter J, Jung S, Keller S et al. Many roads to maturity: microRNA biogenesis pathways and their regulation. Nat Cell Biol 2009; 11(3):228-234.
3. Morlando M, Ballarino M, Gromak N et al. Primary microRNA transcripts are processed cotranscriptionally. Nat Struct Mol Biol 2008; 15(9):902-909.
4. Ballarino M, Pagano F, Girardi E et al. Coupled RNA processing and transcription of intergenic primary microRNAs. Mol Cell Biol 2009; 29(20):5632-5638.
5. Filipowicz W, Bhattacharyya SN, Sonenberg N. Mechanisms of post-transcriptional regulation by microRNAs: are the answers in sight? Nat Rev Genet 2008; 9(2):102-114.
6. Liu X, Fortin K, Mourelatos Z. microRNAs: biogenesis and molecular functions. Brain Pathol 2008; 18(1):113-121.
7. Schmittgen TD. Regulation of microRNA processing in development, differentiation and cancer. J Cell Mol Med 2008; 12(5B):1811-1819.
8. Davis BN, Hilyard AC, Lagna G et al. SMAD proteins control DROSHA-mediated microRNA maturation. Nature 2008; 454(7200):56-61.
9. Fukuda T, Yamagata K, Fujiyama S et al. DEAD-box RNA helicase subunits of the Drosha complex are required for processing of rRNA and a subset of microRNAs. Nat Cell Biol 2007; 9(5):604-611.
10. Guil S, Caceres JF. The multifunctional RNA-binding protein hnRNP A1 is required for processing of miR-18a. Nat Struct Mol Biol 2007; 14(7):591-596.
11. Obernosterer G, Leuschner PJ, Alenius M et al. Post-transcriptional regulation of microRNA expression. Rna 2006; 12(7):1161-1167.
12. Trabucchi M, Briata P, Garcia-Mayoral M et al. The RNA-binding protein KSRP promotes the biogenesis of a subset of microRNAs. Nature 2009; 459(7249):1010-1014.
13. Viswanathan SR, Daley GQ, Gregory RI. Selective blockade of microRNA processing by Lin28. Science 2008; 320(5872):97-100.
14. Suzuki HI, Yamagata K, Sugimoto K et al. Modulation of microRNA processing by p53. Nature 2009; 460(7254):529-533.
15. Yamagata K, Fujiyama S, Ito S et al. Maturation of microRNA is hormonally regulated by a nuclear receptor. Mol Cell 2009; 36(2):340-347.

16. Chen CY, Gherzi R, Ong SE et al. AU binding proteins recruit the exosome to degrade ARE-containing mRNAs. Cell 2001; 107(4):451-464.
17. Garcia-Mayoral MF, Hollingworth D, Masino L et al. The structure of the C-terminal KH domains of KSRP reveals a noncanonical motif important for mRNA degradation. Structure 2007; 15(4):485-498.
18. Gherzi R, Lee KY, Briata P et al. A KH domain RNA binding protein, KSRP, promotes ARE-directed mRNA turnover by recruiting the degradation machinery. Mol Cell 2004; 14(5):571-583.
19. Kroll TT, Zhao WM, Jiang C et al. A homolog of FBP2/KSRP binds to localized mRNAs in Xenopus oocytes. Development 2002; 129(24):5609-5619.
20. Min H, Turck CW, Nikolic JM et al. A new regulatory protein, KSRP, mediates exon inclusion through an intronic splicing enhancer. Genes Dev 1997; 11(8):1023-1036.
21. Ruggiero T, Trabucchi M, De Santa F et al. LPS induces KH-type splicing regulatory protein-dependent processing of microRNA-155 precursors in macrophages. Faseb J 2009; 23(9):2898-2908.
22. Michlewski G, Guil S, Semple CA et al. Post-transcriptional regulation of miRNAs harboring conserved terminal loops. Mol Cell 2008; 32(3):383-393.
23. Ruggiero T, Trabucchi M, Ponassi M et al. Identification of a set of KSRP target transcripts upregulated by PI3K-AKT signaling. BMC Mol Biol 2007; 8:28.
24. Garcia-Mayoral MF, Diaz-Moreno I, Hollingworth D et al. The sequence selectivity of KSRP explains its flexibility in the recognition of the RNA targets. Nucleic Acids Res 2008.
25. Newman MA, Thomson JM, Hammond SM. Lin-28 interaction with the Let-7 precursor loop mediates regulated microRNA processing. Rna 2008; 14(8):1539-1549.
26. Piskounova E, Viswanathan SR, Janas M et al. Determinants of microRNA processing inhibition by the developmentally regulated RNA-binding protein Lin28. J Biol Chem 2008; 283(31):21310-21314.
27. Rybak A, Fuchs H, Smirnova L et al. A feedback loop comprising lin-28 and let-7 controls pre-let-7 maturation during neural stem-cell commitment. Nat Cell Biol 2008; 10(8):987-993.
28. Wulczyn FG, Smirnova L, Rybak A et al. Post-transcriptional regulation of the let-7 microRNA during neural cell specification. Faseb J 2007; 21(2):415-426.
29. Hagan JP, Piskounova E, Gregory RI. Lin28 recruits the TUTase Zcchc11 to inhibit let-7 maturation in mouse embryonic stem cells. Nat Struct Mol Biol 2009; 16(10):1021-1025.
30. Heo I, Joo C, Kim YK et al. TUT4 in concert with Lin28 suppresses microRNA biogenesis through Pre-microRNA uridylation. Cell 2009; 138(4):696-708.
31. Lehrbach NJ, Armisen J, Lightfoot HL et al. LIN-28 and the poly(U) polymerase PUP-2 regulate let-7 microRNA processing in Caenorhabditis elegans. Nat Struct Mol Biol 2009; 16(10):1016-1020.
32. Sakamoto S, Aoki K, Higuchi T et al. The NF90-NF45 complex functions as a negative regulator in the microRNA processing pathway. Mol Cell Biol 2009; 29(13):3754-3769.
33. Chen JF, Mandel EM, Thomson JM et al. The role of microRNA-1 and microRNA-133 in skeletal muscle proliferation and differentiation. Nat Genet 2006; 38(2):228-233.
34. Kim HK, Lee YS, Sivaprasad U et al. Muscle-specific microRNA miR-206 promotes muscle differentiation. J Cell Biol 2006; 174(5):677-687.
35. Allemand E, Guil S, Myers M et al. Regulation of heterogenous nuclear ribonucleoprotein A1 transport by phosphorylation in cells stressed by osmotic shock. Proc Natl Acad Sci USA 2005; 102(10):3605-3610.
36. Shimada N, Rios I, Moran H et al. p38 MAP kinase-dependent regulation of the expression level and subcellular distribution of heterogeneous nuclear ribonucleoprotein A1 and its involvement in cellular senescence in normal human fibroblasts. RNA Biol 2009; 6(3).
37. Briata P, Forcales SV, Ponassi M et al. p38-dependent phosphorylation of the mRNA decay-promoting factor KSRP controls the stability of select myogenic transcripts. Mol Cell 2005; 20(6):891-903.
38. Diaz-Moreno I, Hollingworth D, Frenkiel TA et al. Phosphorylation-mediated unfolding of a KH domain regulates KSRP localization via 14-3-3 binding. Nat Struct Mol Biol 2009; 16(3):238-246.
39. Gherzi R, Trabucchi M, Ponassi M et al. The RNA-binding protein KSRP promotes decay of beta-catenin mRNA and is inactivated by PI3K-AKT signaling. PLoS Biol 2006; 5(1):e5.
40. Thomson JM, Newman M, Parker JS et al. Extensive post-transcriptional regulation of microRNAs and its implications for cancer. Genes Dev 2006; 20(16):2202-2207.
41. Lu J, Getz G, Miska EA et al. microRNA expression profiles classify human cancers. Nature 2005; 435(7043):834-838.
42. Kumar MS, Lu J, Mercer KL. Impaired microRNA processing enhances cellular transformation and tumorigenesis. Nat Genet 2007; 39(5):673-677.
43. Guo J, Li ZC, Feng YH. Expression and activation of the reprogramming transcription factors. Biochem Biophys Res Commun 2009; 390(4):1081-1086.
44. Heo I, Joo C, Cho J et al. Lin28 mediates the terminal uridylation of let-7 precursor microRNA. Mol Cell 2008; 32(2):276-284.
45. Bussing I, Slack FJ, Grosshans H. let-7 microRNAs in development, stem cells and cancer. Trends Mol Med 2008; 14(9):400-409.

CHAPTER 5

HORMONAL REPRESSION OF miRNA BIOSYNTHESIS THROUGH A NUCLEAR STEROID HORMONE RECEPTOR

Sally Fujiyama-Nakamura, Kaoru Yamagata and Shigeaki Kato*

Abstract: The maturation of primary microRNAs (pri-miRNAs) to precursor miRNAs (pre-miRNAs) is mediated by the "microprocessor" complex minimally comprimising two core components, Drosha and DGCR8. However, the roles of RNA-binding proteins associated with these core units in the large Drosha complex remain to be defined. While signal-dependent regulation of miRNA biogenesis is assumed, such regulation remains to be described. Here, we provide a short review based on our recent findings of hormonally-regulated pri-miRNA processing by nuclear estrogen receptor.

INTRODUCTION

miRNAs control cell proliferation/differentiation and fate through modulation of gene expression by partially base-pairing with target mRNA sequences. Mammalian miRNA genes encode noncoding large RNAs and are initially transcribed by RNA polymerase II as mono- or polycistronic precursor pri-miRNAs. Pri-mRNAs are processed into 60~70 nt hairpins with 3'-overhangs by the nuclear RNase III endonuclease Drosha to form pre-miRNAs. Pre-miRNAs translocate into the cytosol and are then processed by Dicer, another RNase III-related enzyme, to generate mature 17-24 nt miRNAs.

In 2004, human Drosha was biochemically identified and found to form two types of complexes.[1] The small complex, named the microprocessor, consists of two subunits, Drosha and DGCR8 (a double-strand RNA-binding protein) and efficiently processes pri-miRNA in vitro. The large complex is comprised of several classes of RNA-binding proteins, including DEAD/DEAH-box type RNA helicases, heterogeneous nuclear

*Corresponding Author: Shigeaki Kato—Institute of Molecular and Cellular Biosciences, University of Tokyo, Yayoi 1-1-1, Bunkyo-ku, Tokyo 113-0032, Japan. Email: uskato@mail.ecc.u-tokyo.ac.jp

Regulation of microRNAs, edited by Helge Großhans.
©2010 Landes Bioscience and Springer Science+Business Media.

ribonucleoproteins (hnRNPs) and putative RNA-binding proteins containing RNA recognition motifs (RRM). However, the roles of these RNA-binding proteins in this large complex have not been clear. Among them, p68/p72 are ATP-dependent DEAD-box RNA helicases. Recently we observed that the p68/p72 components are required for Drosha-mediated processing of specific pri-miRNA substrates in mice.[2] Interestingly, we had previously found that p68/p72 are transcriptional coregulators for nuclear estrogen receptor α (ERα).[3,4] Based upon the previous reports, we tested the hypothesis that estrogen-activated ERα functionally associated with the large Drosha complex through direct interaction with p68/p72 and thereby modulated Drosha-mediated processing of pri-miRNA. Indeed, we found that activated ERα was inhibitory for pri-miRNA processing.[5] In this review, we describe the roles of p68/p72 in nuclear estrogen receptor-mediated miRNA biosynthesis.

p68/p72 DEAD-BOX RNA HELICASES SERVE AS RNA-BINDING COMPONENTS IN THE DROSHA COMPLEX

p68/p72 RNA Helicases in RNA-Related Events

RNA helicases unwind RNA duplexes or disrupt RNA-protein interaction using the energy generated from ATP hydrolysis.[6] Consequently, RNA helicases are assumed to be involved in diverse RNA-related events, including transcription, splicing, translation, ribosome biogenesis, RNA transport, RNA processing and degradation.[6] Of these, p68 (Ddx5) and p72 (Ddx17) are DEAD-box RNA helicases and share 90% amino acid sequence identity in the central core domains, with much less homology at the N- and C-termini[7] (Fig. 1A). p68 and p72 form a heterodimer in cells.[8] p68 was identified as a component of biochemically purified spliceosomal complexes[9,10] and interacts with the U1 snRNA-5' splice site duplex,[11] while p72 was detected in U1 snRNP.[12] These findings suggest p68 and p72 have significant roles in mRNA splicing. Their contribution to ribosomal RNA (rRNA) processing has been demonstrated in yeast.[13,14] p68 also functions as an intercellular signal transducer in β-catenin activation by PDGF, although it is unlikely that its RNA helicase activity is involved.[15] Moreover, transcriptional coregulator activities of p68 and p72 were documented by our previous reports that they co-activate estrogen receptor alpha (ERα) through their direct interaction with the N-terminal A/B domain of ERα.[3,4,16] Thus, diverse roles of p68/p72 RNA helicases in transcription, pre-mRNA and pre-rRNA processing have been documented, presumably reflecting their inherent capacity of associating with various RNAs of distinct tertiary structures.

p68/p72 Are Components of the Large Drosha Complex

Recently, p68/p72 were identified as components of the purified large Drosha complex[1] and shown to be required for processing of rRNA and a subset of pri-miRNAs in mice.[2] p68/p72 appear indispensable for life, since gene disruption of either p68 or p72 in mice results in lethality at birth.[2] Using p72 knockout mice for a global microarray survey of miRNA, we found that the expression of a particular set of miRNAs, but not all 266 surveyed, were down-regulated,[2] suggesting that certain species of miRNAs require p68/p72 for Drosha-mediated processing. A set of overlapping miRNA species was down-regulated in the absence of either p68 or p72, but a few miRNAs (such as

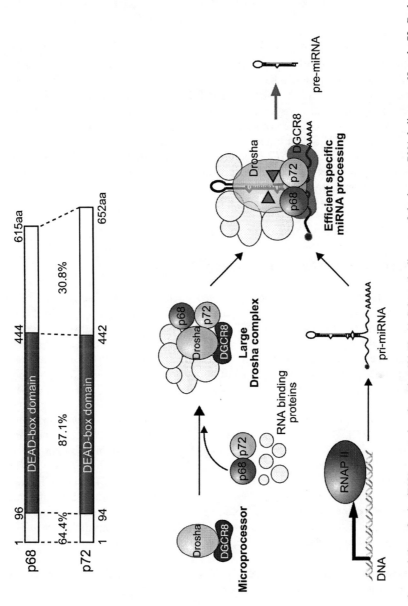

Figure 1. Model for the regulation of pri-miRNA processing through p68/p72. A) Schematic diagram of the human RNA helicases, p68 and p72. Both proteins possess the DEAD-box-type RNA helicase domains in the central regions. The percentage of amino-acid identity is indicated between the two proteins. B) Model for the reaction mechanism of p68 and p72 in Drosha-mediated pri-miRNA processing.

miR-214) appeared to be regulated by only p72, suggesting that p68/p72 helicases have similar but distinct substrate specificities in their recognition of pri-miRNAs. Thus, p68/p72 appear to specifically perceive the structures of RNA substrates and to facilitate cleavage at precise RNA sites by Drosha, presumably owing to their RNA helicase activities (Fig. 1B).

The p68 knockout (KO) in mouse was embryonic lethal by day 11.5, while p72 KO resulted in neonatal death at postnatal day 2. The double knockouts caused earlier lethality than those seen in single KO mice,[2] consistent with previous implications that these proteins are essential for development.[17,18] Likewise, knockout of DGCR8 in mice arrested embryogenesis at early stages and ES cells deficient of DGCR8 were defective in self-renewal.[19] Disorders were also detectable in cultured embryonic fibroblasts deficient in p68 or p72. Taken together, these findings support the idea that proper pri-miRNA processing via p68/p72 is required for cell proliferation and differentiation of embryos. Furthermore, p68/p72 are significant components of the Drosha complex for a subset of pri-miRNAs.

GENE REGULATION BY NUCLEAR ESTROGEN RECEPTORS

Transcriptional Control by Nuclear Estrogen Receptor

The female hormone estrogen participates in diverse physiological actions, including development of female reproductive organs, the central nervous system and bone maintenance.[20-25] The biologically active form of estrogen, 17β-estradiol (E_2), serves as a ligand for its nuclear receptors (ERα and ERβ), which are members of the nuclear steroid hormone receptor superfamily. ERs act as hormone-inducible transcription factors to regulate the expression of a specific set of target genes.[26,27] Homo- or hetero-dimers of ERα and ERβ recognize and stably bind to estrogen response elements (EREs) in target gene promoters.

Like other nuclear receptors, ERα and ERβ consist of several conserved functional domains[28] (Fig. 2A). The N-terminal A/B region is a transcriptional activation domain (AF-1), while the central C region encompasses the DNA-binding domain (DBD) composed of the two cysteine-rich zinc fingers. The E region is bifunctional as the hormone-binding domain (LBD) and the second AF domain (AF-2) and D/F domains are hinge regions. ERα and ERβ share high sequence similarity for the DBD C domain, while the other regions are much less homologous (Fig. 2A). E_2 binding to the LBD induces a conformational alteration of ERs and facilitates recruitments of transcriptional co-activators that are docked to the AF-1 and AF-2 domains.[29]

Transcriptional Activation by Estrogen-Bound Estrogen Receptors Requires Transcriptional Coregulators

Genome-wide ChIP-chip (Chromatin-immunoprecipitation followed by microarray chip analysis) approaches have recently shown that estrogen-bound (activated) ERs associate with chromatin at numerous sites and that FOXA1 (HNF3α), a forkhead family transcription factor, is a pioneer factor that binds to chromatin, opens the chromatin and participates in the recruitment of ERα to several cis-regulatory elements

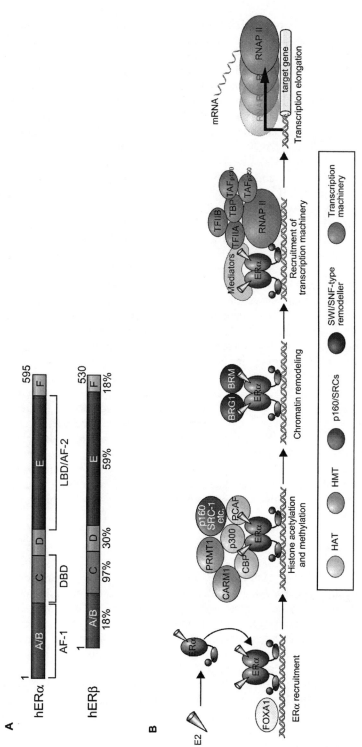

Figure 2. Model for the transcriptional regulatory mechanism by estrogen receptors. A) Schematic representation of the human estrogen receptors, ERα and ERβ. Both receptors consist of six functional domains (regions A to F), including the ligand-independent activation function AF-1, the DNA-binding domain (DBD), the ligand-binding domain (LBD) and the ligand-dependent activation function AF-2. The percentages indicated below ERβ are the amino-acid identities in the regions. B) Model for the transcriptional regulation of estrogen receptor.

on chromatin.[30,31] Chromatin-bound ERs are believed to then recruit a number of transcriptional coregulators for hormone-induced gene activators[26,32,33] (Fig. 2B). Transcriptional coregulators associating with ERs appear to reorganize nucleosomal arrays through histone modification and ATP-dependent chromatin remodeling, which facilitates formation of the transcription initiation complex at promoters.

The best characterized ER AF-2 co-activators are members of the p160 steroid receptor co-activator (SRC) family. The SRC family consists of three 160 kDa proteins: SRC-1 (NCoA-1), SRC-2 (transcriptional intermediary factor 2 (TIF2), GR-interacting protein 1 (GRIP1)) and SRC-3 (ACTR/pCIP/receptor associated coactivator (RAC3)/ TRIM-1/amplified in breast cancer 1 (AIB1)).[34,35] They are common co-activators for many nuclear receptors (NRs) and directly interact with helix 12 in the NR LBDs through their NR box motifs (LxxLL amino acid sequence: L, leucine; x, any amino acid), when helix 12 is shifted by ligand binding to the NR LBDs.[36-38] Besides NR box motifs, p160 proteins bear basic helix-loop-helix (bHLH) domains which are capable of forming complexes with histone acetyltransferases (HAT), such as cAMP-response element binding protein (CREB)-binding protein (CBP) and p300.[36,39,40] The p160 family of proteins docks histone methyltransferases, PRMT1 and CARM1 (PRMT4) to transcriptionally co-activate ERs.[41] Thus, histone acetylation and methylation are likely prerequisite for transcriptional activation by estrogen-bound ERs.[42,43] This idea is also supported by the fact that other HATs and histone methyltransferases are coregulators for nuclear receptors including ERs.[44,45] In response to histone modification, chromatin structure is re-organized by ATP-dependent chromatin remodeling proteins.[46] Indeed, the SWI/SNF-type chromatin remodeling complex is assumed to facilitate ER-mediated transcriptional activation.[47-50]

In contrast to the NR AF-2 co-activators, AF-1 co-activators are believed to be receptor species-selective, since the AF-1 domain is not conserved among NRs. In this regard, p68/p72 were the first co-activators that were identified to be selective for ERα AF-1.[3,4] p68/p72 appear to form a complex with p300/CBP HAT and it was proposed that this bridges the ERα AF-1 and AF-2 domains for efficient transactivation. The co-activator activity of p68/p72 was enhanced when p68/p72 were bound to SRA, a known RNA co-activator.[4,51] Thus, p68/p72 likely constitute a class of ERα AF-1-selective co-activators and their co-activation activity appears to be RNA-dependent, presumably through their RNA helicase activity.

Post-Transcriptional Regulation by Estrogen

In addition to hormonal regulation at the transcriptional level,[27,52,53] estrogen also mediates later regulatory steps. For instance, similar to progesterone androgen and glucocorticoid receptor systems, the estrogen receptor is subject to autoregulatory feedback loops at receptor transcript levels, although the precise mechanism remains to be defined.[54-61] Likewise, estrogen prolongs the half-lives of target gene transcripts such as integrin β3, cyclophilin 4, superoxide dismutase (SOD3), thyrotropin releasing hormone receptor (Trhr) and vascular endothelial growth factor (VEGF).[62-67] As the untranslated regions (UTRs) of mRNA are generally responsible for control of mRNA stability, a number of studies have been performed to delineate the regulatory elements and identify cognate binding proteins which control mRNA stability.[68,69] Nevertheless, studies have thus far failed to fully explain estrogen-mediated stabilization of target mRNAs at a molecular level.

ESTROGEN-INDUCED mRNA STABILITY IS MEDIATED THROUGH HORMONALLY REGULATED miRNA BIOSYNTHESIS

Activated ERα Attenuates Conversion of pri-miRNA into pre-miRNA

As the genes encoding miRNAs are transcribed by RNA polymerase II, we initially reasoned that the promoters of some miRNA genes might harbor EREs. To address this issue, an miRNA microarray analysis was performed in pre-ovariectomized mice following estrogen treatment.[5] Unexpectedly, estrogen treatment down-regulated a set of miRNA in the uterus, while up-regulation was seen only in a few miRNA species. Estrogen-mediated inhibition of miRNA expression in the uterus was also detected in embryos deficient of ERα. In a human breast cancer cell line, MCF7 cells, estrogen was also effective at down-regulating many but not all miRNA species and the estrogen effect could be blocked by knock-down of ERα. Since mature miRNA is generated by Drosha and Dicer in a two-step process, we asked if miRNA processing was attenuated by estrogen-activated ERα. Estrogen did not appear to reduce the expression levels of the pri-miRNA forms of the down-regulated miRNA species. However, down-regulation of pre-miRNAs reflected that seen in miRNAs. Thus, it is most likely at this stage that the promoters of miRNA genes harbor neither regulatory elements for estrogen, nor direct binding sites for ERs. Instead, the processing from pri-miRNA to pre-miRNA appeared to be hormonally regulated by ERα.

Activated ERα Suppresses Processing Activity of Drosha through Its Direct Interaction with p68/p72

Consistent with our previous observations that p68/p72 directly interacted with ERα, activated ERα was found to be capable of associating with p68/p72 in the large Drosha complex. This association was likely to be further potentiated by estrogen binding to ERα, presumably owing to contact of the two LxxLL motifs in Drosha with ERα helix 12. Such estrogen-induced stable association of ERα with the large Drosha complex suppressed the processing of pri-miRNAs into pre-miRNAs in an in vitro processing assay as well as in cultured cells. Direct association of activated ERα with p68/p72 in the large Drosha complex is further supported by in vivo observations that Drosha-mediated processing of a particular set of pri-miRNA species is controlled by either p68/p72 or activated ERα.

Hormonally Down-Regulated miRNAs Stabilize VEGF mRNA in Estrogen-Dependent Breast Cancer Cells

In mammals, miRNA destabilizes mRNA and/or inhibits translation through formation of double-stranded RNA on target UTRs of mRNA.[70-72] As mRNAs of the estrogen target genes are stabilized by estrogen, we investigated the molecular link between the regulated miRNAs and mRNA stability. By searching miRNA target sequences in the UTRs of estrogen target genes mRNAs, human VEGF mRNA was found to bear several target sequences in the 3'UTR for miR-125a and miR-195. These miRNAs were able to destabilize reporter mRNAs containing the target sequences of the 3'UTR in MCF-7 cells. Indeed, hVEGF mRNA expression levels were reduced by the miRNAs targeting the hVEGF 3'UTR. Taken together, we suggest that down-regulation of certain sets of miRNAs results in stabilization of mRNAs of some estrogen target genes (Fig. 3).

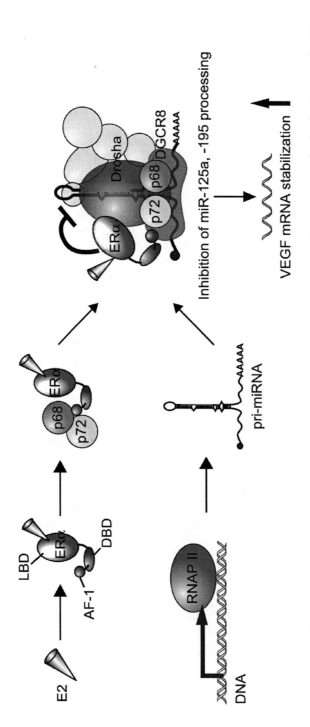

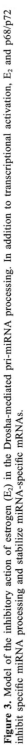

Figure 3. Model of the inhibitory action of estrogen (E₂) in the Drosha-mediated pri-miRNA processing. In addition to transcriptional activation, E₂ and p68/p72 inhibit specific miRNA processing and stabilize miRNA-specific mRNAs.

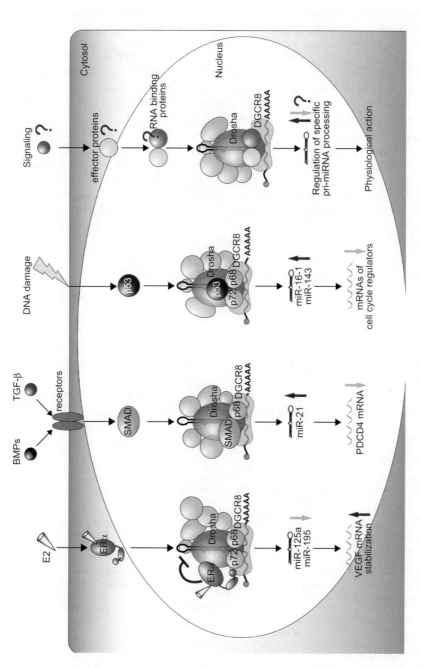

Figure 4. Perspective model for the novel function of RNA-binding proteins in the Drosha-mediated pri-miRNA processing machinery. From the results of the pri-miRNA processing regulations by the stimulation of estrogen,[5] TGF-β signaling[73] and DNA damage,[74] it is conceivable that p68 or other RNA-binding proteins in the large Drosha complex serve as adaptor subunits which cross-talk with other signaling pathways.

CONCLUSION

Estrogen-induced mRNA stabilization has long been documented; however, the molecular basis has remained unclear. Our recent work provides in vivo and in vitro evidence supporting a novel regulatory mechanism in which estrogen attenuates pri-miRNA processing through hormone-dependent association of activated ERα with components of the large Drosha complex and subsequently stabilizes mRNAs otherwise targeted by the miRNA.[5] This hormonal regulation required association of p68/p72 with activated ERα. In addition to ERα, SMAD was recently reported as a direct interactant for p68 in the Drosha complex (see also chapter by Hata and Davis).[73] Interestingly, TGF-β-induced association promoted Drosha-mediated processing of a particular set of miRNAs.[73] More recently, p53 was identified as a p68 interaction partner for efficient processing of a set of pri-miRNAs.[74] Thus, it appears that the RNA helicase p68 has an essential role as a regulatory subunit docking intracellular signal transducers for regulated pri-miRNA processing by the large Drosha complex (Fig. 4). It is still unclear if p72 is equally significant in regulated pri-miRNA processing, but p72 might have a distinct role since the miRNA species regulated by p72 did not fully overlap with those by p68.[2] In this respect, it is conceivable that RNA-binding proteins in the large Drosha complex serve as adaptor subunits which cross-talk with other signaling pathways in addition to facilitating processing of pri-miRNAs with distinct tertiary structures.

ACKNOWLEDGEMENTS

We thank the members of the laboratory of nuclear signaling in IMCB for helpful discussions for this article and H. Yamazaki and M. Yamaki for preparing the manuscript.

Our work has been supported by ERATO, Japan Science and Technology Agency and the Ministry of Science and Higher Education, Grant-in-Aid for Scientific Research.

REFERENCES

1. Gregory RI, Yan KP, Amuthan G et al. The Microprocessor complex mediates the genesis of microRNAs. Nature 2004; 432(7014):235-240.
2. Fukuda T, Yamagata K, Fujiyama S et al. DEAD-box RNA helicase subunits of the Drosha complex are required for processing of rRNA and a subset of microRNAs. Nat Cell Biol 2007; 9(5):604-611.
3. Endoh H, Maruyama K, Masuhiro Y et al. Purification and identification of p68 RNA helicase acting as a transcriptional coactivator specific for the activation function 1 of human estrogen receptor alpha. Mol Cell Biol 1999; 19(8):5363-5372.
4. Watanabe M, Yanagisawa J, Kitagawa H et al. A subfamily of RNA-binding DEAD-box proteins acts as an estrogen receptor alpha coactivator through the N-terminal activation domain (AF-1) with an RNA coactivator, SRA. EMBO J 2001; 20(6):1341-1352.
5. Yamagata K, Fujiyama S, Ito S et al. Maturation of microRNA is hormonally regulated by a nuclear receptor. Mol Cell 2009; 36(2):340-347.
6. Bleichert F, Baserga SJ. The long unwinding road of RNA helicases. Mol Cell 2007; 27(3):339-352.
7. Lamm GM, Nicol SM, Fuller-Pace FV et al. p72: a human nuclear DEAD box protein highly related to p68. Nucleic Acids Res 1996; 24(19):3739-3747.
8. Ogilvie VC, Wilson BJ, Nicol SM et al. The highly related DEAD box RNA helicases p68 and p72 exist as heterodimers in cells. Nucleic Acids Res 2003; 31(5):1470-1480.
9. Neubauer G, King A, Rappsilber J et al. Mass spectrometry and EST-database searching allows characterization of the multi-protein spliceosome complex. Nat Genet 1998; 20(1):46-50.

10. Zhou Z, Licklider LJ, Gygi SP et al. Comprehensive proteomic analysis of the human spliceosome. Nature 2002; 419(6903):182-185.
11. Liu ZR. p68 RNA helicase is an essential human splicing factor that acts at the U1 snRNA-5' splice site duplex. Mol Cell Biol 2002; 22(15):5443-5450.
12. Lee CG. RH70, a bidirectional RNA helicase, copurifies with U1snRNP. J Biol Chem 2002; 277(42):39679-39683.
13. Bond AT, Mangus DA, He F et al. Absence of Dbp2p alters both nonsense-mediated mRNA decay and rRNA processing. Mol Cell Biol 2001; 21(21):7366-7379.
14. Jalal C, Uhlmann-Schiffler H, Stahl H. Redundant role of DEAD box proteins p68 (Ddx5) and p72/p82 (Ddx17) in ribosome biogenesis and cell proliferation. Nucleic Acids Res 2007; 35(11):3590-3601.
15. Yang L, Lin C, Liu ZR. p68 RNA helicase mediates PDGF-induced epithelial mesenchymal transition by displacing Axin from beta-catenin. Cell 2006; 127(1):139-155.
16. Fuller-Pace FV, Ali S. The DEAD box RNA helicases p68 (Ddx5) and p72 (Ddx17): novel transcriptional coregulators. Biochem Soc Trans 2008; 36(Pt 4):609-612.
17. Stevenson RJ, Hamilton SJ, MacCallum DE et al. Expression of the 'dead box' RNA helicase p68 is developmentally and growth regulated and correlates with organ differentiation/maturation in the fetus. J Pathol 1998; 184(4):351-359.
18. Ip FC, Chung SS, Fu WY et al. Developmental and tissue-specific expression of DEAD box protein p72. Neuroreport 2000; 11(3):457-462.
19. Wang Y, Medvid R, Melton C et al. DGCR8 is essential for microRNA biogenesis and silencing of embryonic stem cell self-renewal. Nat Genet 2007; 39(3):380-385.
20. Krege JH, Hodgin JB, Couse JF et al. Generation and reproductive phenotypes of mice lacking estrogen receptor beta. Proc Natl Acad Sci USA 1998; 95(26):15677-15682.
21. Weihua Z, Saji S, Makinen S et al. Estrogen receptor (ER) beta, a modulator of ERalpha in the uterus. Proc Natl Acad Sci USA 2000; 97(11):5936-5941.
22. Couse JF, Korach KS. Estrogen receptor null mice: what have we learned and where will they lead us? Endocr Rev 1999; 20(3):358-417.
23. Bocchinfuso WP, Korach KS. Mammary gland development and tumorigenesis in estrogen receptor knockout mice. J Mammary Gland Biol Neoplasia 1997; 2(4):323-334.
24. Tam J, Danilovich N, Nilsson K et al. Chronic estrogen deficiency leads to molecular aberrations related to neurodegenerative changes in follitropin receptor knockout female mice. Neuroscience 2002; 114(2):493-506.
25. Nakamura T, Imai Y, Matsumoto T et al. Estrogen prevents bone loss via estrogen receptor alpha and induction of Fas ligand in osteoclasts. Cell 2007; 130(5):811-823.
26. Smith CL, O'Malley BW. Coregulator function: a key to understanding tissue specificity of selective receptor modulators. Endocr Rev 2004; 25(1):45-71.
27. Kato S, Sato T, Watanabe T et al. Function of nuclear sex hormone receptors in gene regulation. Cancer Chemother Pharmacol 2005; 56(Suppl 1):4-9.
28. Tsai MJ, O'Malley BW. Molecular mechanisms of action of steroid/thyroid receptor superfamily members. Annu Rev Biochem 1994; 63:451-486.
29. Shiau AK, Barstad D, Loria PM et al. The structural basis of estrogen receptor/coactivator recognition and the antagonism of this interaction by tamoxifen. Cell 1998; 95(7):927-937.
30. Carroll JS, Liu XS, Brodsky AS et al. Chromosome-wide mapping of estrogen receptor binding reveals long-range regulation requiring the forkhead protein FoxA1. Cell 2005; 122(1):33-43.
31. Laganiere J, Deblois G, Lefebvre C et al. From the Cover: Location analysis of estrogen receptor alpha target promoters reveals that FOXA1 defines a domain of the estrogen response. Proc Natl Acad Sci USA 2005; 102(33):11651-11656.
32. McKenna NJ, O'Malley BW. Combinatorial control of gene expression by nuclear receptors and coregulators. Cell 2002; 108(4):465-474.
33. Shibata H, Spencer TE, Onate SA et al. Role of co-activators and corepressors in the mechanism of steroid/thyroid receptor action. Recent Prog Horm Res 1997; 52:141-164; discussion 164-145.
34. McKenna NJ, Lanz RB, O'Malley BW. Nuclear receptor coregulators: cellular and molecular biology. Endocr Rev 1999; 20(3):321-344.
35. Glass CK, Rosenfeld MG. The coregulator exchange in transcriptional functions of nuclear receptors. Genes Dev 2000; 14(2):121-141.
36. Onate SA, Tsai SY, Tsai MJ et al. Sequence and characterization of a coactivator for the steroid hormone receptor superfamily. Science 1995; 270(5240):1354-1357.
37. Heery DM, Kalkhoven E, Hoare S et al. A signature motif in transcriptional co-activators mediates binding to nuclear receptors. Nature 1997; 387(6634):733-736.
38. McInerney EM, Rose DW, Flynn SE et al. Determinants of coactivator LXXLL motif specificity in nuclear receptor transcriptional activation. Genes Dev 1998; 12(21):3357-3368.

39. Kamei Y, Xu L, Heinzel T et al. A CBP integrator complex mediates transcriptional activation and AP-1 inhibition by nuclear receptors. Cell 1996; 85(3):403-414.
40. Torchia J, Glass C, Rosenfeld MG. Co-activators and corepressors in the integration of transcriptional responses. Curr Opin Cell Biol 1998; 10(3):373-383.
41. Xu W, Cho H, Evans RM. Acetylation and methylation in nuclear receptor gene activation. Methods Enzymol 2003; 364:205-223.
42. Allis CD, Berger SL, Cote J et al. New nomenclature for chromatin-modifying enzymes. Cell 2007; 131(4):633-636.
43. Li B, Carey M, Workman JL. The role of chromatin during transcription. Cell 2007; 128(4):707-719.
44. Yang XJ, Ogryzko VV, Nishikawa J et al. A p300/CBP-associated factor that competes with the adenoviral oncoprotein E1A. Nature 1996; 382(6589):319-324.
45. Yanagisawa J, Kitagawa H, Yanagida M et al. Nuclear receptor function requires a TFTC-type histone acetyl transferase complex. Mol Cell 2002; 9(3):553-562.
46. Berger SL. The complex language of chromatin regulation during transcription. Nature 2007; 447(7143):407-412.
47. Ichinose H, Garnier JM, Chambon P et al. Ligand-dependent interaction between the estrogen receptor and the human homologues of SWI2/SNF2. Gene 1997; 188(1):95-100.
48. DiRenzo J, Shang Y, Phelan M et al. BRG-1 is recruited to estrogen-responsive promoters and cooperates with factors involved in histone acetylation. Mol Cell Biol 2000; 20(20):7541-7549.
49. Belandia B, Orford RL, Hurst HC et al. Targeting of SWI/SNF chromatin remodelling complexes to estrogen-responsive genes. Embo J 2002; 21(15):4094-4103.
50. Okada M, Takezawa S, Mezaki Y et al. Switching of chromatin-remodelling complexes for oestrogen receptor-alpha. EMBO Rep 2008; 9(6):563-568.
51. Lanz RB, McKenna NJ, Onate SA et al. A steroid receptor coactivator, SRA, functions as an RNA and is present in an SRC-1 complex. Cell 1999; 97(1):17-27.
52. Mangelsdorf DJ, Thummel C, Beato M et al. The nuclear receptor superfamily: the second decade. Cell 1995; 83(6):835-839.
53. Lonard DM, O'Malley BW. Nuclear receptor coregulators: judges, juries and executioners of cellular regulation. Mol Cell 2007; 27(5):691-700.
54. Mitchell DC, Ing NH. Estradiol stabilizes estrogen receptor messenger ribonucleic acid in sheep endometrium via discrete sequence elements in its 3'-untranslated region. Mol Endocrinol 2003; 17(4):562-574.
55. Flouriot G, Pakdel F, Valotaire Y. Transcriptional and post-transcriptional regulation of rainbow trout estrogen receptor and vitellogenin gene expression. Mol Cell Endocrinol 1996; 124(1-2):173-183.
56. Saceda M, Lindsey RK, Solomon H et al. Estradiol regulates estrogen receptor mRNA stability. J Steroid Biochem Mol Biol 1998; 66(3):113-120.
57. Tseng L, Zhu HH. Regulation of progesterone receptor messenger ribonucleic acid by progestin in human endometrial stromal cells. Biol Reprod 1997; 57(6):1360-1366.
58. Yeap BB, Krueger RG, Leedman PJ. Differential post-transcriptional regulation of androgen receptor gene expression by androgen in prostate and breast cancer cells. Endocrinology 1999; 140(7):3282-3291.
59. Yeap BB, Voon DC, Vivian JP et al. Novel binding of HuR and poly(C)-binding protein to a conserved UC-rich motif within the 3'-untranslated region of the androgen receptor messenger RNA. J Biol Chem 2002; 277(30):27183-27192.
60. Schaaf MJ, Cidlowski JA. Molecular mechanisms of glucocorticoid action and resistance. J Steroid Biochem Mol Biol 2002; 83(1-5):37-48.
61. Ing NH. Steroid hormones regulate gene expression post-transcriptionally by altering the stabilities of messenger RNAs. Biol Reprod 2005; 72(6):1290-1296.
62. Staton JM, Thomson AM, Leedman PJ. Hormonal regulation of mRNA stability and RNA-protein interactions in the pituitary. J Mol Endocrinol. 2000; 25(1):17-34.
63. Li CF, Ross FP, Cao X et al. Estrogen enhances alpha v beta 3 integrin expression by avian osteoclast precursors via stabilization of beta 3 integrin mRNA. Mol Endocrinol 1995; 9(7):805-813.
64. Kumar P, Mark PJ, Ward BK et al. Estradiol-regulated expression of the immunophilins cyclophilin 40 and FKBP52 in MCF-7 breast cancer cells. Biochem Biophys Res Commun 2001; 284(1):219-225.
65. Strehlow K, Rotter S, Wassmann S et al. Modulation of antioxidant enzyme expression and function by estrogen. Circ Res 2003; 93(2):170-177.
66. Kimura N, Arai K, Sahara Y et al. Estradiol transcriptionally and post-transcriptionally up-regulates thyrotropin-releasing hormone receptor messenger ribonucleic acid in rat pituitary cells. Endocrinology 1994; 134(1):432-440.
67. Ruohola JK, Valve EM, Karkkainen MJ et al. Vascular endothelial growth factors are differentially regulated by steroid hormones and antiestrogens in breast cancer cells. Mol Cell Endocrinol 1999; 149(1-2):29-40.

68. Cheadle C, Fan J, Cho-Chung YS et al. Stability regulation of mRNA and the control of gene expression. Ann N Y Acad Sci 2005; 1058:196-204.
69. Kozak M. How strong is the case for regulation of the initiation step of translation by elements at the 3' end of eukaryotic mRNAs? Gene 2004; 343(1):41-54.
70. Zamore PD, Haley B. Ribo-gnome: the big world of small RNAs. Science 2005; 309(5740):1519-1524.
71. Meister G. miRNAs get an early start on translational silencing. Cell 2007; 131(1):25-28.
72. Gu S, Jin L, Zhang F et al. Biological basis for restriction of microRNA targets to the 3' untranslated region in mammalian mRNAs. Nat Struct Mol Biol 2009; 16(2):144-150.
73. Davis BN, Hilyard AC, Lagna G et al. SMAD proteins control DROSHA-mediated microRNA maturation. Nature 2008; 454(7200):56-61.
74. Suzuki HI, Yamagata K, Sugimoto K et al. Modulation of microRNA processing by p53. Nature 2009; 460(7254):529-533.

CHAPTER 6

AUTOREGULATORY MECHANISMS CONTROLLING THE MICROPROCESSOR

Robinson Triboulet and Richard I. Gregory*

Abstract: The Microprocessor, comprising the ribonuclease Drosha and its essential cofactor, the double-stranded RNA-binding protein, DGCR8, is essential for the first step of the miRNA biogenesis pathway. It specifically cleaves double-stranded RNA within stem-loop structures of primary miRNA transcripts (pri-miRNAs) to generate precursor (pre-miRNA) intermediates. Pre-miRNAs are subsequently processed by Dicer to their mature ~22 nt form. Thus, Microprocessor is essential for miRNA maturation, and pri-miRNA cleavage by this complex defines one end of the mature miRNA. Moreover, it is emerging that dysregulation of the Microprocessor is associated with various human diseases. It is therefore important to understand the mechanisms by which the expression of the subunits of the Microprocessor is regulated. Recent findings have uncovered a post-transcriptional mechanism that maintains the integrity of the Microprocessor. These studies revealed that the Microprocessor is involved in the processing of the messenger RNA (mRNA) that encodes DGCR8. This regulatory feedback loop, along with the reported role played by DGCR8 in the stabilization of Drosha protein, is part of a newly identified regulatory mechanism controlling Microprocessor activity.

INTRODUCTION

There are now more than 700 human microRNAs (miRNAs) listed in the miRBase registry http://www.mirbase.org. Each of these is thought to be capable of regulating the expression of hundreds or even thousands of target messenger RNAs (mRNAs).[1] Therefore, miRNA-mediated repression is considered to be an important and pervasive level of gene regulation and is essential for normal development. Apart from a very

*Corresponding Author: Richard I. Gregory—Stem Cell Program, Children's Hospital Boston, Department of Biological Chemistry and Molecular Pharmacology, Harvard Medical School, Harvard Stem Cell Institute Boston, Massachusettes 02115, USA. Email: rgregory@enders.tch.harvard.edu

Regulation of microRNAs, edited by Helge Großhans.
©2010 Landes Bioscience and Springer Science+Business Media.

small number of so-called mirtrons, which are a subgroup of miRNAs found in short introns that bypass the first step of the miRNA biogenesis pathway,[2-4] most mature ~22 miRNAs are sequentially processed by the canonical miRNA-processing pathway.[5] miRNAs are embedded in both coding and noncoding genes, with approximately 40% located in introns and ~10% located in exons for both type of genes. These genes are mainly transcribed by RNA polymerase II.[6,7] The maturation of miRNAs involves the sequential cleavage of primary transcripts called pri-miRNAs that display stem-loop structures containing mature miRNA sequences.[8] The nuclear RNase III enzyme Drosha (also known as RNASEN) was identified as the enzyme responsible for cleavage of the pri-miRNAs to generate hairpin-shaped precursor intermediates (pre-miRNA) of ~60-80 nucleotides in length.[9] Subsequently, it was found that Drosha resides in a multi-subunit complex termed Microprocessor. Microprocessor is a 500-650kDa complex comprising Drosha and its essential cofactor, the double-stranded RNA-binding protein DiGeorge syndrome critical region gene 8 (DGCR8).[10-13] In Drosophila, the DGCR8 homolog is named Pasha (_Partner of Drosha_). In vitro reconstitution experiments have established that DGCR8 is essential for substrate recognition, RNA-binding and recruiting Drosha catalytic activity for pri-miRNA processing.[14] Genetic studies in mice have demonstrated that DGCR8 is essential for normal development and for the rapid proliferation and differentiation of embryonic stem (ES) cells.[15] DGCR8 interacts with the junction between single- and double-stranded RNA at the base of the stem-loop and serves as a ruler for Drosha-mediated endonucleolytic cleavage of the stem 11 base pairs away from the junction, converting pri-miRNA into pre-miRNA.[16] Microprocessor cleaves the double-stranded RNA asymmetrically thereby generating pre-miRNAs with two-nucleotide overhangs at the 3'-end that are ideal substrates for subsequent processing by Dicer in the cell cytoplasm. Processing of pri-miRNA is a cotranscriptional mechanism that is also coupled with pre-mRNA splicing and degradation events.[17-19] Although Microprocessor is necessary and sufficient for the processing of some miRNAs, the biogenesis of a subset of miRNAs is regulated by additional factors, some of which were identified as components of a very large (1-2 MDa) Drosha-containing complex:[10] for example the DEAD-box RNA helicases DDX5 (p68) and DDX17 (p72), which facilitate the processing of some miRNAs,[20] or the RNA-binding protein HNRNPA1, which is required for processing miR-18a from a cluster of miRNAs (miR-17, miR-18a, miR-19a, miR-20a, miR-19b-1 and miR-92) that are generated from a single pri-miRNA transcript (see also chapter by Michlewski et al).[21] Other accessory factors, some of them discussed in other chapters of this book, have also been implicated in pri-miRNA processing including ADAR,[22] SMAD,[23] SNIP1,[24] KSRP,[25] ARS2,[26] p53,[27] the NF90-NF45 complex[28] and ERα[29] although their exact roles in this process remain to be determined.

Dysregulation of miRNA expression is associated with human disease. One of the best characterized examples is the role of altered miRNA expression in cancer.[30] It has been demonstrated that individual miRNAs can act as either tumor suppressors or oncogenes. Interestingly, a global alteration in miRNA expression profiles has been observed in many different primary human tumors. Specifically, it was found that most tumors express a globally lower level of miRNAs than the corresponding normal tissues.[31] Although the mechanism for this phenomenon remains to be determined, DGCR8 mRNA level was modestly changed in those tumors (1.7 fold). In a separate study, it was revealed that the global downregulation of miRNA expression in cancer is post-transcriptional, likely the result of a defect in miRNA processing.[32] Several other studies have observed altered Drosha or DGCR8 expression in various different cancers (see Table 1).[33-40] This raises the intriguing

Table 1. Disease-associated alterations in Microprocessor

Disease	Alteration	Ref.
Cancer		
Various primary human tumors	Global downregulation of miRNA expression in many different human primary tumors. DGCR8 expression decreased	31
Various primary human tumors	miRNA and mRNA profiling revealed that the global downregulation of miRNA expression in human tumors is post-transcriptional	32
Mouse lung adenocarcinoma (LKR13) cells.	DGCR8 or Drosha knockdown in LKR13 cells enhanced cell transformation. In mice, DGCR8/Drosha-depleted cells formed more invasive tumors with accelerated kinetics.	41
Esophageal squamous cell carcinoma	Drosha upregulation. Elevated levels of Drosha associated with poor prognosis. Drosha knockdown in esophageal cancer cell lines reduced proliferation.	33
Cervical squamous cell carcinoma	Drosha upregulation and copy number gain	34
Cervical squamous cell carcinoma	Drosha upregulation and copy number gain	35
Pleomorphic adenomas of the salivary gland	Upregulation of Drosha and DGCR8	39
Ovarian carcinoma	Drosha downregulation. Low Drosha expression associated with poor prognosis	37
Breast	Drosha expressed at lower levels in estrogen receptor negative (ER-) tumors	38
Breast, ovarian and melanoma	Drosha and DGCR8 copy number loss/gain	36
Blood malignancies	Altered DGCR8/Drosha expression	40
Other Disorders		
Del22q11.2 syndrome/DiGeorge Syndrome	Monoallelic microdeletion of chromosome 22q11.2 including the DGCR8 locus	OMIM (NCBI)
Mouse model of Del22q11.2 syndrome/DiGeorge Syndrome	DGCR8 heterozygous deletion. DGCR8 haploinsufficiency in mice contributes to the behavioral and neuronal deficits associated with the 22q11.2 microdeletion.	48
Schizophrenia	Upregulation of DGCR8 expression and global upregulation of miRNA expression in the superior temporal gyrus and the dorsolateral prefrontal cortex	49

hypothesis that perturbation of Microprocessor activity, possibly involving altered DGCR8 and/or Drosha expression, is associated with tumorigenesis. This notion is directly supported by experimental evidence in which shRNA-mediated depletion of DGCR8, Drosha or Dicer enhanced cellular transformation and led to increased tumorigenesis when these knockdown cells were transplanted to recipient mice.[41] Moreover, conditional deletion of Dicer enhanced tumor development in a K-Ras-induced mouse model of lung cancer. Overall, these studies indicate that abrogation of global miRNA processing promotes tumorigenesis.

The gene encoding DGCR8 maps a region of human chromosome 22 that is associated with DiGeorge syndrome. DiGeorge syndrome is the most common human genetic deletion syndrome, with an incidence of around 1:4000 live births and is caused by the monoallelic microdeletion of chromosome 22q11.2 (del22q11.2), with most patients having a typical deleted region (TDR) of ~3Mb, which encompasses >30 genes. The clinical features of the disease are highly variable although the majority of patients are born with congenital heart defects. Other common features include a characteristic facial appearance, immunodeficiency, cleft palate, hypoparathyroidism and developmental and behavioral problems.[42,43] Despite its high incidence, the fact that most cases are caused by heterozygous deletion of a region containing many genes has made the task of identifying the particular genes that underlie the pathogenesis a challenge. Mouse models and the identification of mutations in nondeleted patients, have revealed that haploinsufficiency of the T-box transcription factor, TBX1, causes some of the DiGeorge syndrome phenotypes including heart defects. However, TBX1 is not responsible for other common features including the neurobehavioral phenotype that manifests in early childhood as learning difficulties, cognitive defects and attention-deficit disorder. In adolescence and adulthood, patients often develop various psychiatric disorders including preponderance of schizophrenia.[42,44-47] Therefore other genes located within the TDR must contribute to DiGeorge syndrome. Interestingly, a mouse model revealed that DGCR8 haploinsufficiency is associated with some of the behavioral and neuronal deficits associated with del22q11.2 syndrome.[48] Moreover, miRNA expression profiling revealed a significant increase in global miRNA expression in the brains of schizophrenia patients compared with unaffected individuals.[49] Importantly, a schizophrenia-associated increase in DGCR8 expression was identified in these brain samples. Thus, it is apparent that maintaining the integrity of the Microprocessor is critical for normal development and its dysregulation, which leads to altered miRNA-processing activity, is associated with several diseases including psychological disorders and cancer. Until recently, mechanisms controlling the expression levels of the components of Microprocessor remained unknown.

POST-TRANSCRIPTIONAL REGULATION OF DGCR8 BY THE MICROPROCESSOR

A bioinformatics study using an algorithm called EvoFold, which searches for evolutionary conserved secondary structures, identified that DGCR8 mRNA harbors two hairpin structures, one in the 5' untranslated region (UTR) and another in the coding region close to the start of the open reading frame (ORF).[50] These hairpins are conserved in human, chimpanzee, mouse, rat, dog, chicken, zebra-fish and puffer-fish. A similar observation was reported in Drosophila, where another hairpin-shaped secondary structure was found in the 5'UTR of Pasha mRNA.[51] Remarkably, these structures, which look like pri-miRNAs, are themselves substrates for the Microprocessor and are cleaved in vitro by immunopurified- or recombinant Microprocessor (Fig. 1).[52,53] Processing products resemble pre-miRNA in their size. However both studies noted that the processing of these hairpins is considerably less efficient than seen for the in vitro processing of bona fide pri-miRNAs,

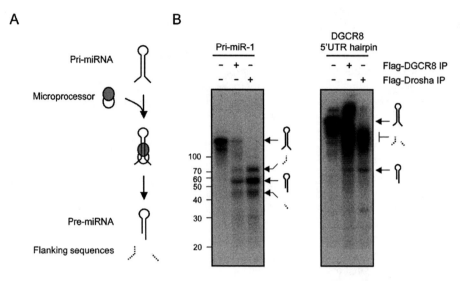

Figure 1. In vitro processing assay of DGCR8 5'UTR RNA by immunopurified Microprocessor. A) Schematic representation of pri-miRNA processing by Microprocessor. B) The indicated in vitro transcribed, internally labelled RNA was incubated with or without Flag-immunopurified DGCR8 (Flag-DGCR8 IP) or Drosha (Flag-Drosha IP). Arrows indicate processing products. RNA was resolved on 15% polyacrylamide denaturing gels and visualized by autoradiography.

suggesting that they do not represent optimal substrates for Microprocessor-mediated cleavage compared to canonical pri-miRNAs. Also, the hairpin in the 5'UTR of DGCR8 seems to be a better substrate for Microprocessor cleavage than the second hairpin located in the beginning of the DGCR8 ORF. Accordingly, the corresponding 'pre-miRNA' from the DGCR8 5'UTR is detected at low levels by Northern blot analysis of RNA extracted from cultured Hela, HEK293 and mouse ES cells. The downstream hairpin in the DGCR8 coding region is undetectable by Northern Blot indicating that this pre-miRNA-like hairpin barely accumulates in cells. The low steady state level of these hairpins may be because either their cleavage is inefficient and/or they are rapidly degraded. Results obtained with a more sensitive PCR-based analysis of DGCR8 mRNA using 5' rapid amplification of cDNA ends (5'RACE) are consistent with cleavage occurring only at the 5'UTR hairpin but not the hairpin located in the DGCR8 coding region. Mature miRNAs of ~22 nt are hardly or not at all detected in cells. However, the 'pre-miRNA' from the 5'UTR can be effectively processed by recombinant Dicer to release a ~22 nt product in vitro. One possible explanation that reconciles the lack of detectable mature miRNA in cells is that the 5'UTR hairpin seems not to be transported to the cell cytoplasm and is therefore inaccessible to Dicer cleavage. Mature miR-1306-3p and miR-1306-5p, which are produced from the second hairpin, have been identified using massive parallel sequencing approaches to analyze small RNA libraries prepared from human ES cells, dog peripheral blood or skin, though again, very few sequences were identified, indicating that these sequences are of very low abundance.[54-56]

Taken together, the aforementioned findings indicate that the hairpin located in the 5'UTR of DGCR8 is a substrate for Microprocessor cleavage both in vitro and in vivo. However, unlike canonical pri-miRNAs that give rise to mature miRNAs for the regulation of target genes, miRNAs are not robustly generated from DGCR8 mRNA. Could this pathway be a novel mechanism that involves the direct post-transcriptional regulation of gene expression by Microprocessor-mediated mRNA cleavage? Indeed, this notion is supported by experiments using either siRNA to deplete Drosha or overexpression of a catalytically inactive mutant Drosha protein. Both depletion of endogenous Drosha or overexpression of mutant Drosha protein is accompanied by an increase in DGCR8 mRNA and protein levels. The destabilization of DGCR8/Pasha by Drosha was also observed in mouse and Drosophila cells, suggesting that this mechanism is evolutionarily conserved.[51-53] Importantly, this post-transcriptional destabilization of DGCR8 mRNA is mediated through Microprocessor cleavage of the 5'UTR hairpin.[52,53] Reporter gene constructs containing DGCR8 5'UTR sequence are stabilized by siRNA-mediated depletion of DGCR8 or Drosha. Thus, the DGCR8 5'UTR confers Microprocessor-dependent repression of a luciferase reporter gene in vivo. Together, these results uncover a novel feedback loop that regulates DGCR8 expression levels.

STABILIZATION OF DROSHA PROTEIN BY DGCR8

An additional level of post-transcriptional control, operating to maintain the expression of Microprocessor subunits, was also uncovered. Although apparently more subtle than the Microprocessor-mediated regulation of DGCR8 mRNA stability, this mechanism seems to involve the stabilization of Drosha protein by association with its binding partner DGCR8. Indeed, depletion of DGCR8 by RNAi is accompanied by a modest decrease of Drosha protein expression whereas Drosha mRNA is not affected.[52] Conversely, DGCR8 overexpression results in accumulation of Drosha protein, but mutants of DGCR8 that do not to interact with Drosha fail to stabilize it. A similar observation is reported in DGCR8[-/-] mouse embryonic (ES) cells, where Drosha protein levels are lower compared to DGCR8[flox/flox] (wild-type) cells. Remarkably, in heterozygous DGCR8[-/flox] ES cells, a compensatory phenomenon is observed. In these cells, DGCR8 expression should be reduced by 50% and consequently Drosha protein expression should also be reduced compared to wild-type cells. Actually, Drosha and DGCR8 mRNA and protein are expressed at similar levels as in wild-type cells. Moreover, Microprocessor activity seems to be quite equivalent in both types of cells, where the level of pri-miRNA and corresponding mature miRNA is almost unchanged (over 85% of those in wild-type ES cells).[15,52] This suggests that these post-transcriptional mechanisms operate to regulate DGCR8 and Drosha expression to maintain a certain level of Microprocessor and miRNA in the cell. The stabilization of other proteins involved in the miRNA and siRNA pathways has been reported previously. For example, the levels of Dicer and its binding partner, the double-stranded RNA-binding partner, TRBP, seem to be interdependent.[57,58] Similarly, Dicer-2 and R2D2 proteins require each other to be stabilized in Drosophila.[59] This seems to be a more general theme in RNA metabolic pathways involving RNA-binding proteins and interacting partners, as exemplified by another recent study showing cross regulation between two factors involved in nonsense-mediated decay.[60]

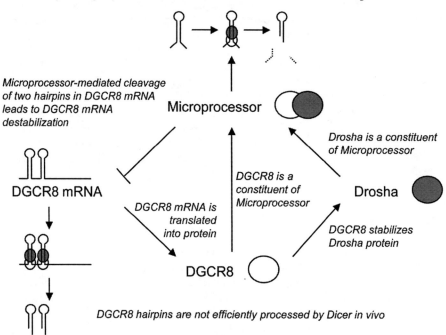

Microprocessor processes pri-miRNA to yield pre-miRNA
By binding to pri-miRNA, DGCR8 assists Drosha-mediated cleavage of the stem

Microprocessor-mediated cleavage
of two hairpins in DGCR8 mRNA
leads to DGCR8 mRNA
destabilization

Microprocessor

Drosha is a constituent
of Microprocessor

DGCR8 mRNA

DGCR8 is a
constituent of
Microprocessor

Drosha

DGCR8 mRNA is
translated
into protein

DGCR8 stabilizes
Drosha protein

DGCR8

DGCR8 hairpins are not efficiently processed by Dicer in vivo

Figure 2. Model for autoregulatory mechanisms controlling the Microprocessor. DGCR8 contains two double-stranded RNA-binding domains that endow Microprocessor with the ability to recognize pri-miRNAs and to guide Drosha-mediated cleavage of the stem of pri-miRNA hairpins. Microprocessor also directly cleaves the hairpin in the 5'UTR of DGCR8 mRNA leading to mRNA destabilization and decreased DGCR8 expression. This feedback mechanism helps to maintain a steady-state level of DGCR8 expression to control the subunit stoichiometry of the Microprocessor, which is physiologically required to achieve efficient processing of pri-miRNA. Drosha protein is stabilized through its interaction with DGCR8.

CONCLUSION

Accurate regulation of Drosha and DGCR8 expression is important to maintain the correct Microprocessor abundance, subunit stoichiometry and miRNA homeostasis. The expression of both proteins is regulated posttranscriptionally by different mechanisms (Fig. 2). DGCR8 mRNA contains two hairpin structures that are recognized and processed by the Microprocessor. The Microprocessor-mediated negative feedback mechanism controls DGCR8 expression by destabilization of DGCR8 mRNA. Of the two hairpins in the DGCR8 mRNA, the 5'UTR hairpin is processed more efficiently in vitro and is detectable by Northern blot analysis of RNA extracted from cells. Indeed, the 5'UTR hairpin alone is sufficient to confer Microprocessor-dependent regulation to a luciferase reporter construct. Therefore the autoregulatory control of DGCR8 expression involves primarily and perhaps exclusively, the 5'UTR hairpin. Given the high degree of sequence and structural conservation in different organisms it seems likely that the hairpin in the

DGCR8 coding region is functionally important for regulating some aspect of DGCR8 expression. Interestingly, an RNA-affinity purification strategy recently identified several proteins that specifically associate with this hairpin in Hela cell extracts.[61] Of note were several ribosomal proteins as well as other RNA-binding proteins implicated in translation control, indicating that this hairpin may have a distinct role in controlling DGCR8 expression. Future work will likely illuminate the exact role of this hairpin and its associated proteins. The autoregulation by an RNAse III enzyme seems to be a convenient and more widespread gene regulatory mechanism. In *E. coli*, the bacterial RNase III, rnc, has been shown to cleave its own mRNA at sites in the 5'-noncoding region of the rnc operon leading to RNA destabilization.[62] It will be interesting to investigate the role of these regulations in primary tumors where a block in Microprocessor activity has been reported, or other instances where expression levels for these proteins are changed.[31,32] DGCR8 mRNA is the first coding transcript to be identified that is processed and destabilized by the Microprocessor. This raises the possibility that other mRNAs could be targeted by Microprocessor endonucleolytic activity. Several studies indicate that hundreds of protein-coding mRNAs could be under the negative regulation of Microprocessor.[51,52,63] These transcripts, which are found to be downregulated in a Microprocessor-dependent and miRNA-independent manner in microarray analyses, are good candidates for direct regulation by Microprocessor cleavage. However, deep-sequencing analyses looking for 18 to 200-nt processing products generated by Microprocessor-mediated cleavage suggested that DGCR8 mRNA is the only mRNA targeted by the Microprocessor.[64] We cannot rule out yet the possibility that processing of mRNA could be cell-type or tissue-specific, or yield highly unstable and/or larger processing products. Therefore, an important open question is whether this Microprocessor-mediated mRNA cleavage mechanism is dedicated to the regulation of DGCR8 expression or whether this pathway is more generally utilized.

PERSPECTIVES

Since both Microprocessor subunits are essential for pri-miRNA processing activity and since correct complex stoichiometry is required for efficient processing, it is likely that there are additional regulatory mechanisms that help maintain the subunit composition of this complex. Many questions currently remain unanswered and future work is likely to provide significant insight into the different modes of regulation. For example, what is the role of different possible posttranslational modifications in the control of Drosha and/or DGCR8 expression and what is the identity of these modifying enzymes? Indeed, proteomic studies have identified phospho-peptides corresponding to both Drosha[65-67] and DGCR8[68-70] proteins. Although the functional significance of these phosphorylation modifications remains to be determined, it was recently demonstrated that the Dicer-binding partner TRBP is phosphorylated by the mitogen-activated protein kinase (MAPK) Erk and TRBP phosphorylation enhanced miRNA production by increasing stability of the Dicer-TRBP complex.[58] Do cell-signaling pathways regulate Microprocessor activity? A possible link between DGCR8 and heme-mediated signal transduction pathway has been proposed since DGCR8 is a heme-binding protein.[71] Also, although the signaling events for this phenomenon are presently obscure, it has been observed that miRNA biogenesis is globally activated in confluent cell cultures.[72] This increase in miRNA expression seems to result from enhanced processing of pri-miRNAs by Microprocessor. Is Microprocessor developmentally regulated? Since tumorigenesis often involves the de-differentiation of cancer cells and the reestablishment of a more embryonic gene expression signature, it is possible that DGCR8 and Drosha expression and thus Microprocessor activity may be developmentally regulated. Indeed, we have found changes in DGCR8 and Drosha expression associated with cell differentiation status (Triboulet and Gregory unpublished data) and upregulation of DGCR8 mRNA occurs during the reprogramming of fibroblasts to ES cell-like induced pluripotent stem (iPS) cells.[73] Can we identify small molecules to manipulate Microprocessor activity? Although so far no high-throughput screens have been performed to directly monitor Microprocessor activity, it seems plausible that such a screen will identify compounds capable of modulating miRNA biogenesis. A small molecule screen for activators of RNAi identified a compound (Enoxacin) that can also enhance miRNA processing, an effect dependent on TRBP.[74]

In summary, recent studies have uncovered novel mechanisms controlling the expression of DGCR8 and Drosha. Indeed, the identification of these and other novel regulatory pathways controlling Microprocessor may lead to the development of novel therapeutics for the treatment of human diseases that are associated with the global dysregulation of miRNA expression including cancer and certain psychiatric disorders.

ACKNOWLEDGEMENTS

R.I.G is supported by lab start-up funds from The Children's Hospital Boston and grants from the NIH-NIGMS: (1R01GM086386-01A1), the Harvard Stem Cell Institute, the March of Dimes Basil O'Connor award and the Emerald Foundation. R.I.G is a Pew Research Scholar.

REFERENCES

1. Bartel DP. microRNAs: target recognition and regulatory functions. Cell 2009; 136:215-33.
2. Ruby JG, Jan CH, Bartel DP. Intronic microRNA precursors that bypass Drosha processing. Nature 2007; 448:83-6.
3. Okamura K, Hagen JW, Duan H et al. The mirtron pathway generates microRNA-class regulatory RNAs in Drosophila. Cell 2007; 130:89-100.
4. Berezikov E, Chung WJ, Willis J et al. Mammalian mirtron genes. Mol Cell 2007; 28:328-36.
5. Kim VN, Han J, Siomi MC. Biogenesis of small RNAs in animals. Nat Rev Mol Cell Biol 2009; 10:126-39.
6. Lee Y, Kim M, Han J et al. microRNA genes are transcribed by RNA polymerase II. EMBO J 2004; 23:4051-60.
7. Cai X, Hagedorn CH, Cullen BR. Human microRNAs are processed from capped, polyadenylated transcripts that can also function as mRNAs. RNA 2004; 10:1957-66.
8. Lee Y, Jeon K, Lee JT et al. microRNA maturation: stepwise processing and subcellular localization. EMBO J 2002; 21:4663-70.
9. Lee Y, Ahn C, Han J et al. The nuclear RNase III Drosha initiates microRNA processing. Nature 2003; 425:415-9.
10. Gregory RI, Yan KP, Amuthan G et al. The Microprocessor complex mediates the genesis of microRNAs. Nature 2004; 432:235-40.
11. Denli AM, Tops BB, Plasterk RH et al. Processing of primary microRNAs by the Microprocessor complex. Nature 2004; 432:231-5.
12. Han J, Lee Y, Yeom KH et al. The Drosha-DGCR8 complex in primary microRNA processing. Genes Dev 2004; 18:3016-27.
13. Landthaler M, Yalcin A, Tuschl T. The human DiGeorge syndrome critical region gene 8 and its D. melanogaster homolog are required for miRNA biogenesis. Curr Biol 2004; 14:2162-7.

14. Yeom KH, Lee Y, Han J et al. Characterization of DGCR8/Pasha, the essential cofactor for Drosha in primary miRNA processing. Nucleic Acids Res 2006; 34:4622-9.
15. Wang Y, Medvid R, Melton C et al. DGCR8 is essential for microRNA biogenesis and silencing of embryonic stem cell self-renewal. Nat Genet 2007; 39:380-5.
16. Han J, Lee Y, Yeom KH et al. Molecular basis for the recognition of primary microRNAs by the Drosha-DGCR8 complex. Cell 2006; 125:887-901.
17. Kim YK, Kim VN. Processing of intronic microRNAs. EMBO J 2007; 26:775-83.
18. Morlando M, Ballarino M, Gromak N et al. Primary microRNA transcripts are processed cotranscriptionally. Nat Struct Mol Biol 2008; 15:902-9.
19. Pawlicki JM, Steitz JA. Primary microRNA transcript retention at sites of transcription leads to enhanced microRNA production. J Cell Biol 2008; 182:61-76.
20. Fukuda T, Yamagata K, Fujiyama S et al. DEAD-box RNA helicase subunits of the Drosha complex are required for processing of rRNA and a subset of microRNAs. Nat Cell Biol 2007; 9:604-11.
21. Guil S, Cáceres JF. The multifunctional RNA-binding protein hnRNP A1 is required for processing of miR-18a. Nat Struct Mol Biol 2007; 14:591-6.
22. Yang W, Chendrimada TP, Wang Q et al. Modulation of microRNA processing and expression through RNA editing by ADAR deaminases. Nat Struct Mol Biol 2006; 13:13-21.
23. Davis BN, Hilyard AC, Lagna G et al. SMAD proteins control DROSHA-mediated microRNA maturation. Nature 2008; 454:56-61.
24. Yu B, Bi L, Zheng B et al. The FHA domain proteins DAWDLE in Arabidopsis and SNIP1 in humans act in small RNA biogenesis. Proc Natl Acad Sci USA 2008; 105:10073-8.
25. Trabucchi M, Briata P, Garcia-Mayoral M et al. The RNA-binding protein KSRP promotes the biogenesis of a subset of microRNAs. Nature 2009; 459:1010-4.
26. Gruber JJ, Zatechka DS, Sabin LR et al. Ars2 links the nuclear cap-binding complex to RNA interference and cell proliferation. Cell 2009; 138:328-39.
27. Suzuki HI, Yamagata K, Sugimoto K et al. Modulation of microRNA processing by p53. Nature 2009; 460:529-33.
28. Sakamoto S, Aoki K, Higuchi T et al. The NF90-NF45 complex functions as a negative regulator in the microRNA processing pathway. Mol Cell Biol 2009; 29:3754-69.
29. Yamagata K, Fujiyama S, Ito S et al. Maturation of microRNA is hormonally regulated by a nuclear receptor. Mol Cell 2009; 36:340-7.
30. Croce CM. Causes and consequences of microRNA dysregulation in cancer. Nat Rev Genet 2009; 10:704-14.
31. Lu J, Getz G, Miska EA et al. microRNA expression profiles classify human cancers. Nature 2005; 435:834-8.
32. Thomson JM, Newman M, Parker JS et al. Extensive post-transcriptional regulation of microRNAs and its implications for cancer. Genes Dev 2006; 20:2202-7.
33. Sugito N, Ishiguro H, Kuwabara Y et al. RNASEN regulates cell proliferation and affects survival in esophageal cancer patients. Clin Cancer Res 2006; 12:7322-8.
34. Muralidhar B, Goldstein LD, Ng G et al. Global microRNA profiles in cervical squamous cell carcinoma depend on Drosha expression levels. J Pathol 2007; 212:368-77.
35. Scotto L, Narayan G, Nandula SV et al. Integrative genomics analysis of chromosome 5p gain in cervical cancer reveals target over-expressed genes, including Drosha. Mol Cancer 2008; 7:58.
36. Zhang X, Cairns M, Rose B et al. Alterations in miRNA processing and expression in pleomorphic adenomas of the salivary gland. Int J Cancer 2009; 124:2855-63.
37. Merritt WM, Lin YG, Han LY et al. Dicer, Drosha and outcomes in patients with ovarian cancer. N Engl J Med 2008; 359:2641-50.
38. Blenkiron C, Goldstein LD, Thorne NP et al. microRNA expression profiling of human breast cancer identifies new markers of tumor subtype. Genome Biol 2007; 8:R214.
39. Zhang L, Huang J, Yang N et al. microRNAs exhibit high frequency genomic alterations in human cancer. Proc Natl Acad Sci USA 2006; 103:9136-41.
40. Lawrie CH, Cooper CD, Ballabio E et al. Aberrant expression of microRNA biosynthetic pathway components is a common feature of haematological malignancy. Br J Haematol 2009; 145:545-8.
41. Kumar MS, Lu J, Mercer KL et al. Impaired microRNA processing enhances cellular transformation and tumorigenesis. Nat Genet 2007; 39(5):673-7.
42. Lindsay EA. Chromosomal microdeletions: dissecting del22q11 syndrome. Nat Rev Genet 2001; 2:858-68.
43. Yamagishi H, Srivastava D. Unraveling the genetic and developmental mysteries of 22q11 deletion syndrome. Trends Mol Med 2003; 9:383-9.
44. Lindsay EA, Vitelli F, Su H et al. Tbx1 haploinsufficieny in the DiGeorge syndrome region causes aortic arch defects in mice. Nature 2001; 410:97-101.

45. Merscher S, Funke B, Epstein JA et al. TBX1 is responsible for cardiovascular defects in velo-cardio-facial/ DiGeorge syndrome. Cell 2001; 104:619-29.
46. Jerome LA, Papaioannou VE. DiGeorge syndrome phenotype in mice mutant for the T-box gene, Tbx1. Nat Genet 2001; 27:286-91.
47. Yagi H, Furutani Y, Hamada H et al. Role of TBX1 in human del22q11.2 syndrome. Lancet 2003; 362:1366-73.
48. Stark KL, Xu B, Bagchi A et al. Altered brain microRNA biogenesis contributes to phenotypic deficits in a 22q11-deletion mouse model. Nat Genet 2008; 40:751-60.
49. Beveridge NJ, Gardiner E, Carroll AP et al. Schizophrenia is associated with an increase in cortical microRNA biogenesis. Mol Psychiatry 2009; Epub ahead of print.
50. Pedersen JS, Bejerano G, Siepel A et al. Identification and classification of conserved RNA secondary structures in the human genome. PLoS Comput Biol 2006; 2:e33.
51. Kadener S, Rodriguez J, Abruzzi KC et al. Genome-wide identification of targets of the drosha-pasha/ DGCR8 complex. RNA 2009; 15:537-45.
52. Han J, Pedersen JS, Kwon SC et al. Post-transcriptional crossregulation between Drosha and DGCR8. Cell 2009; 136:75-84.
53. Triboulet R, Chang HM, Lapierre RJ et al. Post-transcriptional control of DGCR8 expression by the Microprocessor. RNA 2009; 15:1005-11.
54. Morin RD, O'Connor MD, Griffith M et al. Application of massively parallel sequencing to microRNA profiling and discovery in human embryonic stem cells. Genome Res 2008; 18:610-21.
55. Friedländer MR, Chen W, Adamidi C et al. Discovering microRNAs from deep sequencing data using miRDeep. Nat Biotechnol 2008; 26:407-15.
56. Yi R, Pasolli HA, Landthaler M et al. DGCR8-dependent microRNA biogenesis is essential for skin development. Proc Natl Acad Sci USA 2009; 106:498-502.
57. Lee Y, Hur I, Park SY et al. The role of PACT in the RNA silencing pathway. EMBO J 2006; 25:522-32.
58. Paroo Z, Ye X, Chen S et al. Phosphorylation of the human microRNA-generating complex mediates MAPK/Erk signaling. Cell 2009; 139:112-22.
59. Liu Q, Rand TA, Kalidas S et al. R2D2, a bridge between the initiation and effector steps of the Drosophila RNAi pathway. Science 2003; 301:1921-5.
60. Chan WK, Bhalla AD, Le Hir H et al. A UPF3-mediated regulatory switch that maintains RNA surveillance. Nat Struct Mol Biol 2009; 16:747-53.
61. Butter F, Scheibe M, Mörl M et al. Unbiased RNA-protein interaction screen by quantitative proteomics. Proc Natl Acad Sci USA 2009; 106:10626-31.
62. Bardwell JC, Régnier P, Chen SM et al. Autoregulation of RNase III operon by mRNA processing. EMBO J 1989; 8:3401-7.
63. Ganesan G, Rao SM. A novel noncoding RNA processed by Drosha is restricted to nucleus in mouse. RNA 2008; 14:1399-410.
64. Shenoy A, Blelloch R. Genomic analysis suggests that mRNA destabilization by the microprocessor is specialized for the auto-regulation of Dgcr8. PLoS One 2009; 4:e6971.
65. Olsen JV, Blagoev B, Gnad F et al. Global, in vivo and site-specific phosphorylation dynamics in signaling networks. Cell 2006; 127:635-48.
66. Rikova K, Guo A, Zeng Q et al. Global survey of phosphotyrosine signaling identifies oncogenic kinases in lung cancer. Cell 2007; 131:1190-203.
67. Sui S, Wang J, Yang B et al. Phosphoproteome analysis of the human Chang liver cells using SCX and a complementary mass spectrometric strategy. Proteomics 2008; 8:2024-34.
68. Cantin GT, Yi W, Lu B et al. Combining protein-based IMAC, peptide-based IMAC and MudPIT for efficient phosphoproteomic analysis. J Proteome Res 2008; 7:1346-51.
69. Dephoure N, Zhou C, Villén J et al. A quantitative atlas of mitotic phosphorylation. Proc Natl Acad Sci USA 2008; 105:10762-7.
70. Chen RQ, Yang QK, Lu BW et al. CDC25B mediates rapamycin-induced oncogenic responses in cancer cells. Cancer Res 2009; 69:2663-8.
71. Faller M, Matsunaga M, Yin S et al. Heme is involved in microRNA processing. Nat Struct Mol Biol 2007; 14:23-9.
72. Hwang HW, Wentzel EA, Mendell JT. Cell-cell contact globally activates microRNA biogenesis. Proc Natl Acad Sci USA 2009; 106:7016-21.
73. Sridharan R, Tchieu J, Mason MJ et al. Role of the murine reprogramming factors in the induction of pluripotency. Cell 2009; 136:364-77.
74. Shan G, Li Y, Zhang J et al. A small molecule enhances RNA interference and promotes microRNA processing. Nat Biotechnol 2008; 26:933-40.

CHAPTER 7

REGULATION OF pre-miRNA PROCESSING

Nicolas J. Lehrbach and Eric A. Miska*

Abstract: microRNAs are endogenously expressed ~21 nucleotide noncoding RNAs. microRNA-mediated regulation of the translation of specific mRNA is implicated in a range of developmental processes and pathologies. As such, miRNA expression is tightly controlled in normal development by both transcriptional and post-transcriptional mechanisms. This chapter is concerned with the control of pre-miRNA processing of individual miRNAs by specific factors. It is focussed on the regulation of a subset of miRNAs by the RNA-binding protein Lin28/LIN-28. We discuss how Lin28/LIN-28 can sequester pre-let-7 miRNA precursor to prevent Dicer-mediated processing. We describe how interaction of pre-let-7 with Lin28/LIN-28 leads to pre-let-7 uridylation and subsequent degradation. Finally, we analyze how let-7 and Lin28/LIN-28 together act as a highly conserved developmental switch that controls stem cell differentiation in *C. elegans* and mammals.

INTRODUCTION

Precise spatial and temporal expression of specific microRNAs is essential to ensure their targets are repressed appropriately. Selective regulation of the processing of individual miRNAs at the level of the pre-miRNA might have distinct advantages. First, regulating the immediate precursor might provide a rapid-response control mechanism over mature miRNA levels. Second, selective recognition might be easier to achieve for pre-miRNAs than for pri-miRNAs as the former have a more tightly defined feature set, a single hairpin structure of 70-90 bases in length. Indeed, a number of examples of specific miRNA regulation at the level of pre-miRNA processing have been reported. In 2006 the laboratory of Javier Martinez described differential expression of mature miRNAs and their direct precursors in mammals.[1] This study demonstrated by northern

*Corresponding Author: Eric A. Miska—Wellcome Trust Cancer Research UK Gurdon Institute and Department of Biochemistry, University of Cambridge, The Henry Wellcome Building of Cancer and Developmental Biology, Tennis Court Rd, Cambridge, CB2 1QN, UK. Email: e.miska@gurdon.cam.ac.uk

Regulation of microRNAs, edited by Helge Großhans.
©2010 Landes Bioscience and Springer Science+Business Media.

blotting and in situ hybridization that mature miR-138 is spatially restricted to distinct cell types, while its precursor, pre-miR-138-2, is ubiquitously expressed throughout all tissues analyzed. Furthermore, pre-miR-138-2 was found to be exported from the nucleus to the cytoplasm, suggesting that cleavage of this pre-miRNA by Dicer is restricted to certain tissues and cell types. However, the factors controlling pre-miR-138-2 processing remain unknown. More recently, the RNA-binding protein KSRP/KHSRP was reported to promote the biogenesis of a subset of miRNAs in mammalian cells.[2] KSRP/KHSRP was found to be associated with a Dicer-containing complex and to promote the maturation of pre-let-7a-1, pre-miR-20, pre-miR-23b and pre-miR-26b. For more information on KSRP/KHSRP see the chapter by Trabucchi et al. in this book.

The most studied example of selective regulation of pre-miRNA processing to date is centered around the RNA-binding protein Lin28/LIN-28.[3,4] This review will focus on this protein, its miRNA targets, its mechanism of action and its biology. For simplicity we will refer to mammalian LIN-28 orthologs as Lin28.

miRNAs AND DEVELOPMENTAL TIMING IN *C. ELEGANS*

Interest in the genes controlling developmental timing in *C. elegans*[5-7] led to the cloning of the first miRNA, *lin-4* miRNA,[8] and the identification of the first miRNA target, *lin-14* mRNA.[9] The developmental-timing, or heterochronic, pathway regulates stage-specific processes during *C. elegans* larval development. For a detailed review of this pathway, please see Rougvie et al.[10] One focus of the study of the heterochronic pathway in *C. elegans* has been the developmental fate of several stem cells in the lateral hypodermis, collectively known as the seam cells. The seam cells undergo a cell division pattern that is synchronised with the four larval molts of the animal (Fig. 1). Only at the adult stage will the seam cells exit mitosis and terminally differentiate. In *lin-4* mutant animals, the seam cells repeat the cell division pattern that characterises the first larval stage (L1) and fail to differentiate. This mutant phenotype has been interpreted as a heterochronic change with the developmental clock being stuck at the L1 stage, resulting in developmental "retardation". Gain-of-function mutations in the *lin-4* miRNA target *lin-14* lead to the identical phenotype, whereas loss-of-function mutations in *lin-14* result in an opposite, "precocious" phenotype, where the seam cells skip the cell division of the first larval stage. The *lin-4* and *lin-14* gene products therefore act as a developmental switch that controls the L1 to L2 transition.

Three miRNAs of the *let-7* family, *mir-48*, *mir-84* and *mir-241* act redundantly to control the next developmental transition.[11-14] Loss-of-function mutations in these three microRNAs lead to the repetition of the cell division pattern of the second larval stage, whereas a gain-of-function mutation in *mir-48* results in a precocious phenotype. A likely target of *mir-48*, *mir-84* and *mir-241* during this transition is the *C. elegans hunchback* orthologue *hbl-1*. The miRNA *let-7*, the second miRNA to be identified,[14] controls the transition from the fourth larval stage to the adult stage and two of its targets in the heterochronic pathway are the *lin-41* and *hbl-1* genes, both of which are also heterochronic genes.[15-17] More recently, two additional *let-7* target genes, the transcription factors *daf-12* and *pha-4*, were identified.[18] *daf-12* is also a regulator of the heterochronic pathway controlling seam cell fate.[19] Finally, the two *let-7* family microRNAs *mir-48* and *mir-84* also control the cessation of the larval molting cycle at the adult stage, with *mir-48*; *mir-84* double mutant animals undergoing a supernumerary molt at the adult stage.[11]

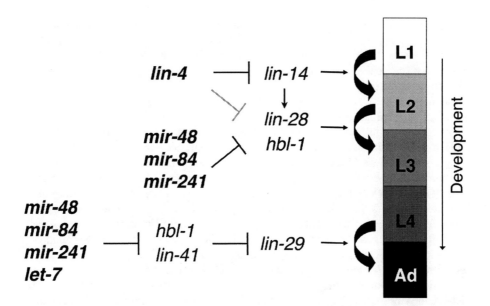

Figure 1. The *let-7* family of miRNAs and the *lin-4* miRNA regulate timing of larval development in *C. elegans*. miRNAs including *let-7* were discovered in *C. elegans* through the analysis of mutants defective in the timing of cell divisions during larval development. *C. elegans* development comprises four larval stages, L1, L2, L3 and L4, each separated by a molt. L4 larvae develop into adults (Ad) through a final molt. The miRNA *lin-4* and the *let-7* family miRNAs *mir-48*, *mir-84* and *mir-241* are required early in development to control appropriate transition between the L1 and L2 stage and the L2 and L3 stage, respectively. The *let-7* miRNA and *mir-48*, *mir-84* and *mir-241* are required for the transition from the L4 to the adult stage. The link between the early larval transitions and the *let-7*-dependent L4 to adult transitions were unclear until recently. microRNAs are in bold, target genes in normal font.

It is striking that at least two microRNA families and at least four microRNAs are involved in the control of developmental timing in *C. elegans*. As the *lin-4* and *let-7* microRNA families are conserved, they might play similar roles in other organisms. This notion is supported by the temporal regulation of *let-7* expression in several species.[20] However, at least one potential role for *let-7* family microRNAs outside the heterochronic pathway has been reported, as discussed below.[21] The microRNAs in the developmental timing pathway act to precisely control temporal expression of their target genes to ensure stable developmental transitions. However, the mechanisms that ensure the precise temporal expression patterns of the microRNAs themselves have largely remained elusive.

THE HETEROCHRONIC GENE *lin-28* ENCODES A REGULATOR OF *let-7* MICRORNA PROCESSING

Lin28/LIN-28 is a conserved cytoplasmic RNA-binding protein containing cold-shock and CHCC Zinc-finger domains.[3,4] In *C. elegans*, LIN-28 is expressed during L1 and L2 stages, but is down regulated during later development, in a *lin-4*-dependent fashion.[3] Mutants lacking *lin-28* are precocious; seam cells perform an L3 specific pattern of cell division during the L2 stage. Animals bearing a transgene causing *lin-28* gain of function display a retarded phenotype similar to that of *lin-4* loss-of-function or *lin-14* gain-of-function mutants. Until recently, the targets through which *lin-28* acts to specify temporal cell fates were unknown. However, the observation that *let-7* mutations almost completely suppress the precocious phenotype of *lin-28* mutants suggested that *lin-28* might act by regulating this miRNA.[14]

The mouse LIN-28 orthologue Lin28 blocks the processing of let-7 in embryonic stem (ES) cells.[22] Initial reports suggested that Lin28 blocks processing of let-7 primary transcripts, based on in vitro Drosha-mediated processing assays.[22,23] However, subsequent studies have shown that mouse Lin28 can similarly prevent Dicer-mediated processing.[24,25] In *C. elegans,* LIN-28 blocks let-7 processing at the Dicer step in vivo.[26] As such, regulation of Dicer-mediated processing is the conserved and likely major mechanism by which Lin28/LIN-28 regulates *let-7*. Both nematodes and mammals have multiple let-7 family miRNAs. In *C. elegans*, LIN-28 regulates *let-7*, but not other *let-7* family miRNAs during development and the functions of *let-7* are distinct from those of the *let-7* family miRNAs (as discussed above). It is not known whether different *let-7* family miRNAs have distinct functions in mice and humans, but loss of Lin28 results in elevated levels of many let-7 family miRNAs and proposed determinants of Lin28 recognition are shared throughout the family (see below). So, it is possible that Lin28 could regulate all let-7 family pre-miRNAs in these species. Most vertebrate studies have focused on single let-7 family members, but the observations made are likely to apply to other let-7 family miRNAs as well.

Consistent with a role in regulation of pre-let-7 processing by Dicer, Lin28/LIN-28 directly binds to pre-let-7.[22-26] The precise mechanism by which Lin28 discriminates pre-let-7 from other cellular RNAs is not completely understood. Various studies have indicated that Lin28 binds pre-let-7 by specifically recognizing the terminal loop of the let-7 pre-miRNA hairpin (Fig. 2).[23,25,27,28] Piskounova et al. found a conserved cytosine residue in the loop of pre-let-7g, which is required for Lin28 binding in vitro. In contrast, Heo et al. identified a GGAG motif in the loop of pre-let-7. The motif is required for Lin28 binding in vitro and is also sufficient to induce binding when introduced into the loop

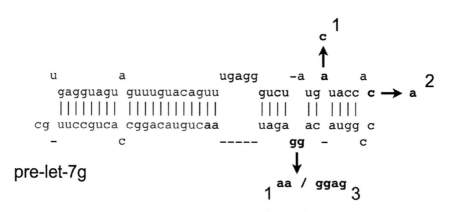

Figure 2. Mutations in the loop-region of pre-let-7 abolish Lin28 binding. Model of the predicted secondary structure of human pre-let-7g. Bases corresponding to mature let-7g (5′ arm) and let-7g* (3′ arm) are highlighted. Bases that have been shown to disrupt interaction with Lin28 in vitro when mutated are shown in bold. 1, data as described.[23] 2, data as described.[22] 3, data were obtained for pre-let-7a and Lin28b as described.[27]

-of unrelated pre-miRNAs. This motif is partially overlapping with sequences required for Lin28 blockade of Drosha-mediated processing in Newman et al. Interestingly, the GGAG motif is found in some additional unrelated miRNAs and these can also be bound by Lin28 in vitro.[27] The terminal loop of pre-let-7 is not conserved between *C. elegans* and vertebrates, yet mouse and human LIN-28 orthologues can still bind *C. elegans* pre-let-7 in vitro[25,29] (E.A.M. and H. Lightfoot, unpublished results), suggesting some conservation in the mechanism of recognition. Interestingly, Rybak et al. found that RNAs that correspond to the let-7 mature miRNA sequence, or to the loop of pre-let-7 can compete for Lin28 binding to pre-let-7. More detailed studies are needed to resolve this. The conserved domain structure of Lin28/LIN-28 is unique among animal proteins. Both the cold shock and zinc finger domains may potentially interact with RNA and so could mediate interaction with pre-let-7, but how this is achieved is currently poorly understood, although both domains appear to contribute to binding and blockade of processing in in vitro (cell-free) assays and in cell culture.[28,29]

Lin28/LIN-28 PROMOTES URIDYLATION AND DEGRADATION OF PRE-LET-7

Once bound to pre-let-7 how does Lin28/LIN-28 prevent it from being processed by Dicer? One possibility is that Lin28 binds to the precursor and physically prevents Dicer from processing it. This suggestion is supported by the fact that Lin28 can prevent processing of pre-let-7 in in vitro assays using recombinant Dicer.[25] However it is contradicted by the observation that steady state levels of pre-let-7 are not altered by perturbation of *Lin28/lin-28*,[23,24,26] indicating that Lin28/LIN-28 might inhibit processing as well as promote turnover of pre-let-7 by another factor.

A study by Heo et al found that pre-let-7 is terminally uridylated in a Lin28-dependent fashion.[24] The uridylated form of pre-let-7 cannot be processed by Dicer and instead is subject to rapid decay by a Dicer-independent mechanism that does not produce mature miRNA. This supports the notion that rather than simply blocking processing, Lin28/ LIN-28 mediates turnover of pre-let-7 molecules. Subsequently, three independent studies identified the enzyme responsible both in humans (TUT4) and *C. elegans* (PUP-2).[26,27,30] TUT4/PUP-2 is a member of a family of poly(U) polymerases, which add U residues to the 3' termini of RNA in a substrate-independent fashion.[31] A direct interaction between TUT4/PUP-2 and Lin28/LIN-28 serves to recruit the enzyme to specifically uridylate pre-let-7. In *C. elegans* the phenotypic consequences of *pup-2* loss of function are much less severe than for *lin-28*, suggesting that *pup-2* enhances, but is not essential for *lin-28*-mediated blockade of *let-7* processing. It will be interesting to compare the roles of let-7 family miRNAs, Lin28 and TUT4 in vivo in vertebrates. A current model of the interactions of pre-let-7, Lin28/LIN-28 and TUT4/PUP-2 is shown in Figure 3.

HETEROCHRONIC GENE ORTHOLOGUES: ANCIENT STEM CELL REGULATORS?

The let-7 miRNA is remarkable in that the entire 21 nucleotide sequence of the mature miRNA is perfectly conserved in a wide range of species including nematodes and man; in addition temporal regulation of let-7 is conserved in many bilateria

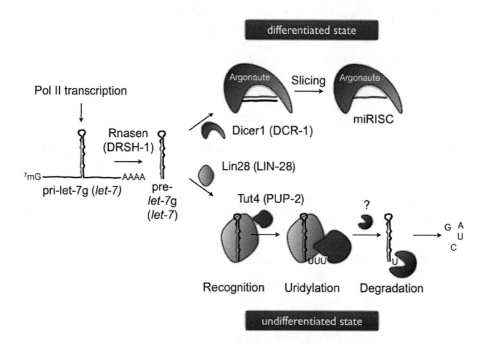

Figure 3. A proposed model for pre-let-7 regulation by Lin28 and a poly(U) polymerase. Lin28(LIN-28) sequesters pre-let-7 in the cytoplasm to prevent Dicer-dependent processing. However, pre-let-7 does not accumulate, but is modified by 3′ uridylation by Tut4(PUP-2) and targeted for degradation. Lin28(LIN-28) and let-7 can act as a switch that regulates cell differentiation. This pathway is conserved from nematodes to mammals. A switch from Lin28(LIN-28) to let-7 expression is reinforced through direct negative regulation of Lin28(LIN-28) by let-7 (not shown). Mouse and *C. elegans* nomenclature are shown.

including sea urchins (Fig. 4).[20] As in *C. elegans*, mature let-7 is absent at early stages of development and highly expressed in mature tissues, while pre-let-7 is expressed throughout development. It is tempting to speculate that temporal expression of let-7 is regulated by lin-28 orthologs throughout the bilateria.

Regulation of let-7 by lin-28 provides a direct link between the early and late developmental timing pathways in *C. elegans* (Fig. 1) and provides the mechanism by which let-7 accumulation is controlled to ensure appropriate timing of differentiation in development. The deep conservation of this regulatory relationship suggests this may be a more universal mechanism that serves to control decisions between growth and maturation in animals. Indeed, Lin28 has been linked in genome-wide association studies to traits such as growth, onset of puberty and menopause.[32-35]

This may reflect a role for Lin28/LIN-28 and let-7 in regulating the decision between proliferation and differentiation at the cellular level. Lin28/LIN-28 and let-7 are reciprocally expressed in stem cells and differentiated tissues; Lin28/LIN-28 levels are high in stem cells, whereas let-7 levels are high in differentiated cells and misexpression of either can be used to drive cell fates in vitro.[36,37] Indeed, Lin28 is one of a small set of factors which when introduced into primary human or mouse cell cultures can promote 'reprogramming' of these cells to induced pluripotent stem (iPS) cells.[37] Two

A

B

Strongylocentrotus purpuratus

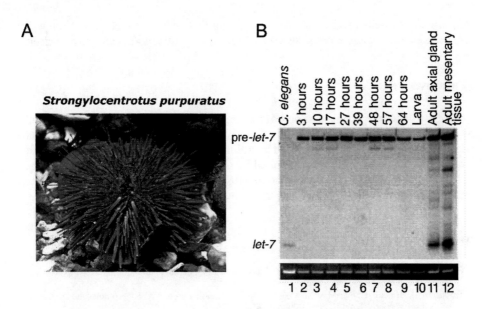

pre-*let-7*

let-7

1 2 3 4 5 6 7 8 9 10 11 12

Figure 4. Developmental regulation of pre-let-7 is an evolutionary conserved mechanism. A) A photograph of the sea urchin *Strongylocentrotus purpuratus* (Kirt L. Onthank). B) In the sea urchin *S. purpuratus* pre-let-7 processing is developmentally regulated.[20] Northern blot for let-7. Lane 1, total RNA from mixed-stage *C. elegans*. Lanes 2-12, total RNA from *S. purpuratus* isolated at different stages of development.

transcription factors, Oct3/4 and Sox2, are necessary and sufficient for iPS formation, but reprogramming efficiency is improved by factors including Lin28, Nanog, c-Myc and Klf4.[37,38] iPS cells can be maintained indefinitely in culture and bear morphological and molecular hallmarks of embryonic stem (ES) cells.[39] In the mouse, iPS cells, like ES cells can contribute to all germ layers in chimeric animals generated after injection into blastocyst stage embryos.[40,41] Consistent with the notion that Lin28 promotes iPS formation by preventing let-7 expression, let-7 inhibitors improve efficiency of iPS induction in the absence of Lin28.[42] Conversely, the overexpression of let-7 can inhibit self-renewal of ES cells; this is likely through downregulation of pluripotency factors including Lin28 and c-Myc.

Finally, deregulation of the let-7/Lin28 switch has been associated with human cancer.[24,36,43,44] Let-7 can act as a tumour suppressor; direct targets of let-7 include the oncogenes Ras, HMGA2 and Lin28 itself.[45] let-7 is absent from a wide range of cancers and introduction of let-7 can reduce tumorigenicity in mouse models.[46] Lin28, in contrast, can act as an oncogene, is highly expressed in many cancers and overexpression of Lin28 can drive the transformation of human cells.[36,43,44]

CONCLUSION

Beginning with studies of the timing of developmental decisions in *C. elegans*, Lin28/ LIN-28 and let-7 have been uncovered as conserved regulators that control proliferation in human stem cells and cancer. Future studies of the functions of Lin28 and let-7 not only promise to yield insights into the mechanisms that underlie development, but will also be relevant to applications in stem cell-based therapeutics and treatment of cancer.

ACKNOWLEDGEMENTS

Figure 4a was reproduced as part of a creative commons license and was authored by Kirt L. Onthank. Figure 4b was reproduced with kind permission from Gary Ruvkun (Harvard/MGH). N.J.L. was supported by a European Framework 6 Integrated Project Grant (SIROCCO). E.A.M. was supported by Cancer Research UK Programme Grant (C13474) and core funding to the Wellcome Trust/Cancer Research UK Gurdon Institute provided by the Wellcome Trust (UK) and Cancer Research UK.

REFERENCES

1. Obernosterer G, Leuschner PJF, Alenius M et al. Post-transcriptional regulation of microRNA expression. RNA 2006; 12(7):1161-1167.
2. Trabucchi M, Briata P, Garcia-Mayoral M et al. The RNA-binding protein KSRP promotes the biogenesis of a subset of microRNAs. Nature 2009; 459(7249):1010-1014.
3. Moss EG, Lee RC, Ambros V. The cold shock domain protein LIN-28 controls developmental timing in C. elegans and is regulated by the lin-4 RNA. Cell 1997; 88(5):637-646.
4. Moss EG, Tang L. Conservation of the heterochronic regulator Lin-28, its developmental expression and microRNA complementary sites. Dev Biol 2003; 258(2):432-442.
5. Ambros V, Horvitz HR. Heterochronic mutants of the nematode Caenorhabditis elegans. Science 1984; 226(4673):409-416.
6. Chalfie M, Horvitz HR, Sulston JE. Mutations that lead to reiterations in the cell lineages of C. elegans. Cell 1981; 24(1):59-69.
7. Horvitz HR, Sulston JE. Isolation and genetic characterization of cell-lineage mutants of the nematode Caenorhabditis elegans. Genetics 1980; 96(2):435-454.
8. Lee RC, Feinbaum RL, Ambros V. The C. elegans heterochronic gene lin-4 encodes small RNAs with antisense complementarity to lin-14. Cell 1993; 75(5):843-854.
9. Wightman B, Ha I, Ruvkun G. Post-transcriptional regulation of the heterochronic gene lin-14 by lin-4 mediates temporal pattern formation in C. elegans. Cell 1993; 75(5):855-862.
10. Rougvie AE. Intrinsic and extrinsic regulators of developmental timing: from miRNAs to nutritional cues. Development 2005; 132(17):3787-3798.
11. Abbott AL, Alvarez-Saavedra E, Miska EA et al. The let-7 microRNA family members mir-48, mir-84 and mir-241 function together to regulate developmental timing in Caenorhabditis elegans. Dev Cell 2005; 9(3):403-414.
12. Lau NC, Lim LP, Weinstein EG et al. An abundant class of tiny RNAs with probable regulatory roles in Caenorhabditis elegans. Science 2001; 294(5543):858-862.
13. Li M, Jones-Rhoades MW, Lau NC et al. Regulatory mutations of mir-48, a C. elegans let-7 family microRNA, cause developmental timing defects. Dev Cell 2005; 9(3):415-422.
14. Reinhart BJ, Slack FJ, Basson M et al. The 21-nucleotide let-7 RNA regulates developmental timing in Caenorhabditis elegans. Nature 2000; 403(6772):901-906.
15. Abrahante JE, Daul AL, Li M et al. The Caenorhabditis elegans hunchback-like gene lin-57/hbl-1 controls developmental time and is regulated by microRNAs. Dev Cell 2003; 4(5):625-637.
16. Lin S-Y, Johnson SM, Abraham M et al. The C elegans hunchback homolog, hbl-1, controls temporal patterning and is a probable microRNA target. Dev Cell 2003; 4(5):639-650.
17. Slack FJ, Basson M, Liu Z et al. The lin-41 RBCC gene acts in the C. elegans heterochronic pathway between the let-7 regulatory RNA and the LIN-29 transcription factor. Mol Cell 2000; 5(4):659-669.
18. Grosshans H, Johnson T, Reinert KL et al. The temporal patterning microRNA let-7 regulates several transcription factors at the larval to adult transition in C. elegans. Dev Cell 2005; 8(3):321-330.
19. Antebi A, Culotti JG, Hedgecock EM. daf-12 regulates developmental age and the dauer alternative in Caenorhabditis elegans. Development 1998; 125(7):1191-1205.
20. Pasquinelli AE, Reinhart BJ, Slack F et al. Conservation of the sequence and temporal expression of let-7 heterochronic regulatory RNA. Nature 2000; 408(6808):86-89.

21. Johnson SM, Grosshans H, Shingara J et al. RAS is regulated by the let-7 microRNA family. Cell 2005; 120(5):635-647.
22. Viswanathan SR, Daley GQ, Gregory RI. Selective blockade of microRNA processing by Lin28. Science 2008; 320(5872):97-100.
23. Newman MA, Thomson JM, Hammond SM. Lin-28 interaction with the Let-7 precursor loop mediates regulated microRNA processing. RNA 2008; 14(8):1539-1549.
24. Heo I, Joo C, Cho J et al. Lin28 mediates the terminal uridylation of let-7 precursor microRNA. Mol Cell 2008; 32(2):276-284.
25. Rybak A, Fuchs H, Smirnova L et al. A feedback loop comprising lin-28 and let-7 controls pre-let-7 maturation during neural stem-cell commitment. Nat Cell Biol 2008; 10(8):987-993.
26. Lehrbach N, Armisen J, Lightfoot H et al. LIN-28 and the poly(U) polymerase PUP-2 regulate let-7 microRNA processing in Caenorhabditis elegans. Nat Struct Mol Biol 2009; 10:1016-1020.
27. Heo I, Joo C, Kim Y-K et al. TUT4 in concert with Lin28 suppresses microRNA biogenesis through pre-microRNA uridylation. Cell 2009; 138(4):696-708.
28. Piskounova E, Viswanathan SR, Janas M et al. Determinants of microRNA processing inhibition by the developmentally regulated RNA-binding protein Lin28. J Biol Chem 2008; 283(31):21310-21314.
29. Balzer E, Heine C, Jiang Q et al. LIN28 alters cell fate succession and acts independently of the let-7 microRNA during neurogliogenesis in vitro. Development 2010; 137(6):891-900.
30. Hagan JP, Piskounova E, Gregory RI. Lin28 recruits the TUTase Zcchc11 to inhibit let-7 maturation in mouse embryonic stem cells. Nat Struct Mol Biol 2009; 16(10):1021-1025.
31. Kwak JE, Wickens M. A family of poly(U) polymerases. RNA 2007; 13(6):860-867.
32. He LN, Recker RR, Deng HW et al. A polymorphism of apolipoprotein E (APOE) gene is associated with age at natural menopause in Caucasian females. Maturitas 2009; 62(1):37-41.
33. Ong KK, Elks CE, Li S et al. Genetic variation in LIN28B is associated with the timing of puberty. Nat Genet 2009; 41:729-733.
34. Perry JR, Stolk L, Franceschini N et al. Meta-analysis of genome-wide association data identifies two loci influencing age at menarche. Nat Genet 2009; 41:648-650.
35. Sulem P, Gudbjartsson DF, Rafnar T et al. Genome-wide association study identifies sequence variants on 6q21 associated with age at menarche. Nat Genet 2009; 41:734-738.
36. West JA, Viswanathan SR, Yabuuchi A et al. A role for Lin28 in primordial germ-cell development and germ-cell malignancy. Nature 2009; 460(7257):909-913.
37. Yu F, Yao H, Zhu P et al. let-7 regulates self renewal and tumorigenicity of breast cancer cells. Cell 2007; 131(6):1109-1123.
38. Wernig M, Meissner A, Cassady JP et al. c-Myc is dispensable for direct reprogramming of mouse fibroblasts. Cell Stem Cell 2008; 2(1):10-12.
39. Takahashi K, Yamanaka S. Induction of pluripotent stem cells from mouse embryonic and adult fibroblast cultures by defined factors. Cell 2006; 126(4):663-676.
40. Wernig M, Meissner A, Foreman R et al. In vitro reprogramming of fibroblasts into a pluripotent ES-cell-like state. Nature 2007; 448(7151):318-324.
41. Maherali N, Sridharan R, Xie W et al. Directly reprogrammed fibroblasts show global epigenetic remodeling and widespread tissue contribution. Cell Stem Cell 2007; 1(1):55-70.
42. Melton C, Judson RL, Blelloch R. Opposing microRNA families regulate self-renewal in mouse embryonic stem cells. Nature 2010; 463:621-426.
43. Chang T-C, Zeitels LR, Hwang HW et al. Lin-28B transactivation is necessary for Myc-mediated let-7 repression and proliferation. Proc Natl Acad Sci USA 2009; 106(9):3384-3389.
44. Viswanathan SR, Powers JT, Einhorn W et al. Lin28 promotes transformation and is associated with advanced human malignancies. Nat Genet 2009; 41(7):843-848.
45. Johnson CD, Esquela-Kerscher A, Stefani G et al. The let-7 microRNA represses cell proliferation pathways in human cells. Cancer Res 2007; 67(16):7713-7722.
46. Yu J, Vodyanik MA, Smuga-Otto K et al. Induced pluripotent stem cell lines derived from human somatic cells. Science 2007; 318(5858):1917-1920.

CHAPTER 8

THE EFFECT OF RNA EDITING AND ADARs ON miRNA BIOGENESIS AND FUNCTION

Bret S.E. Heale, Liam P. Keegan and Mary A. O'Connell*

Abstract: From analysis of deep-sequencing data it is apparent that sequence differences occur between the genome and miRNAs. Changes from genomic A to an apparent G in miRNA can be accounted for by the editing activity of ADARs. Questions that arise from this observation are: How many miRNAs are edited and to what frequency? Is there a specific step in the biogenesis of miRNAs that is preferentially susceptible to editing by ADARs? However the key question is whether editing affects the downstream activity of miRNAs. Despite much evidence that miRNAs are edited, critical examination of the functional data shows a dearth of examples where editing has been demonstrated to actually affect the downstream miRNA activity in vivo. Even where it is demonstrated that RNA editing can affect biogenesis or targeting of a particular miRNA, effects may be limited by redundancy within the miRNA network.

INTRODUCTION

The ADARs (adenosine deaminases acting on RNA) are the main family of RNA editing enzymes found in mammals. They convert specific adenosines to inosines in double-stranded (ds) RNA (Fig. 1). It is primarily the RNA duplex structure that is recognised, rather than the sequence (for review ref. 1). Inosine is read as if it were guanosine by the translation machinery so that editing within coding sequences can result in another amino acid being inserted at the editing site. This can have a huge impact on the properties of the encoded protein. For example, when transcripts encoding subunit B of the AMPA class of a glutamate-gated ion channel receptor are edited so that glutamine (Q) is converted into arginine (R) by ADAR2 at Q/R site, this receptor

*Corresponding Author: Mary A. O'Connell—MRC Human Genetics Unit, Institute of Genetics and Molecular Medicine, Western General Hospital, Crewe Road, Edinburgh EH4 2XU, UK.
Email: M.O'Connell@hgu.mrc.ac.uk

Regulation of microRNAs, edited by Helge Großhans.
©2010 Landes Bioscience and Springer Science+Business Media.

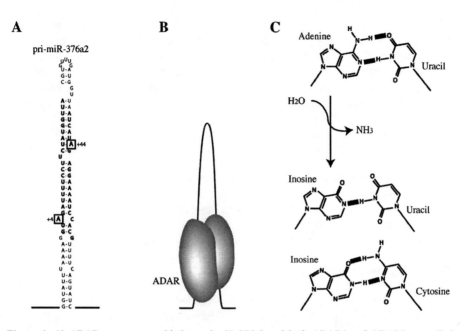

Figure 1. A) ADAR enzymes can bind to pri-miR-376a2 and both ADAR1 and ADAR2 can edit it at the +4 and +44 positions, the mature miRNAs are in bold. B) ADAR proteins recognise duplex RNA and bind to it as a dimer. C) ADAR enzymes can convert adenosine to inosine via hydrolytic deamination so that inosine now base-pairs with cytosine.

is subsequently impermeable to calcium.[2] In most of the transcripts that are specifically edited by ADARs at one or two positions, editing results in recoding and most of these site-specifically edited transcripts are expressed in the CNS. However RNA duplexes formed from opposite-sense transcripts of repetitive elements are also edited. This editing is very prevalent within the human genome due to the abundance of *Alu* elements with the capacity to form RNA hairpins within transcript introns and UTR regions.[3] Yet despite the prevalence of this type of editing its biological function is unknown and as yet can only be speculated on.

When the *Drosophila* and human genomes were first sequenced it came as a surprise that they contained more proteins with dsRNA-binding domains (dsRBD) than anticipated.[4,5] At that time it was thought that there was very little dsRNA in the cell so why have so many proteins with dsRBDs? Amongst the known proteins with dsRBDs at the time were ADAR1, which has three dsRBDs and a catalytic deaminase domain in the carboxy terminus in addition to ADAR2, ADAR3 and TENR which all have two dsRBDs. So far, only ADAR1 and ADAR2 have been demonstrated to be active whereas no catalytic activity has been shown for ADAR3 or TENR. The ADAR1p110 isoform is a nuclear protein found constitutively in most tissue. The ADAR1 p150 isoform is expressed from an interferon-inducible promoter; this protein shuttles in and out of the nucleus and accumulates mainly in the cytoplasm.[6] ADAR2 is a nuclear protein expressed in most tissues but at a lower level than ADAR1 whereas ADAR3 is brain-specific and TENR is only found in the testis.

Considering that dsRNA is in the A form, the view has been that proteins that bind dsRNA do so in a rather nonspecific way with the structure rather than the sequence being important. The major groove in the A form duplex is too narrow and inaccessible to amino acid side chains and the minor grove, while more accessible, has fewer different hydrogen bonding determinants on the bases that would allow sequence-specific recognition.[7] Consequently binding of dsRBDs to dsRNA was considered to be rather nonspecific with specific contacts to the nucleotides not being very important.[8] When it emerged that many of the proteins with dsRBDs were involved in the processing of miRNAs the assumption was that ADARs would either enhance or antagonise this pathway since they bind to a very similar substrate. So the search began to identify the interactions between ADAR proteins and the miRNA pathway and determine at what steps these occur. As ADAR proteins are present both in the nucleus and the cytoplasm, they could potentially influence both the nuclear and cytoplasmic processing of miRNA and their downstream functions.

There may be RNA editing of nucleosides other than adenosine. High-throughput sequencing studies of miRNAs have revealed not only A to G changes and poly-uridylation, but also G-to-A, C-to-U, U-to-C and U-to-A changes. There have even been suggestions of insertions and deletions of nucleotides.[9] In plants, there is evidence of a protective effect of adenylation of miRNAs[10] and evidence that 2'-O-methylation protects miRNAs from 3'-end uridylation and 3'-to-5' exonuclease-mediated degradation.[11,12] Further, in animals Lin-28 has been demonstrated to be involved in uridylation of pre-let-7 as a step in RNA degradation.[13] Liver specific miR-122 undergoes adenylation by the cytoplasmic poly(A) polymerase GLD-2, which stabilizes it and prevents uridylation and subsequently degradation.[14] However in mammals, the most widespread and best studied example of RNA editing of miRNA is adenosine to inosine (A-to-I) deamination by ADARs. Consequently, the main question addressed in this chapter is: what is the evidence that RNA editing by ADARs affects the miRNA pathway?

PREVALENCE OF EDITED miRNAs

When investigating the effect of A-to-I editing on miRNA biogenesis one first has to determine how many miRNA are edited and to what extent the individual miRNAs are edited. Addressing this question is not straightforward as ADAR editing of either the pri-miRNA transcript or the pre-miRNA can also result in inhibition of the production of mature miRNA, potentially reducing the number of detectably edited miRNAs.

An additional consideration is the tissue expression profile of ADARs, which is high in the central nervous system (CNS) but low in tissues such as muscle. Most of the transcripts that are specifically edited are found in the CNS and the complexity of this tissue may not be well represented in sequence data. Also many cell lines, even though they express ADAR proteins, do not have high endogenous editing activity; this is in particular true for ADAR2. Comparison of neuronal tissue and neuronal cell lines might therefore reveal conflicting results. Also, in most cases authors of miRNA cloning and discovery projects avoid analyzing sequences that do not match to the genome. Understandably, they are attempting to limit the noise of their sequencing data, but this has the side effect that editing events go unrecorded. Yet despite these caveats some authors have directly investigated the prevalence and abundance of A-to-I editing in miRNA while others have revealed insight through high-throughput cloning screens.

Perhaps the best data on pri-miRNA editing in humans is from Blow and colleagues who examined editing of pri-miRNAs and estimated that 6% of mature miRNAs are edited.[15] They chose 231 pri-miRNA transcripts to investigate in adult human brain, heart, liver, lung, ovary, placenta, skeletal muscle, small intestine, spleen and testis. Of these 99 were amplified and sequenced and they found that twelve pri-miRNAs were edited, with six of these having editing within the mature miRNA sequence. The editing frequency at single sites ranged from 10% to 70% and was tissue-specific. For example, pri-miR-376a (position 49) was edited 70% in brain and 0% in skeletal muscle. These results are a strong indicator that pri-miRNA can be substrates for A-to-I editing. However, the analysis excludes the investigation of pre-miRNA editing and is not a comprehensive examination of all miRNAs.

Landgraf and colleagues state in the miRNA Expression Atlas, that they found A-to-I editing in about 2% of mature miRNA clones.[16] The consortium's robust data was derived from collections of clones of small RNAs and miRNA arrays. Their analysis covered over 140 samples from several tissue types and cell lines. They even produced libraries from cell lines infected with virus. The primary goal of the authors was to determine abundance and presence of mature miRNA in different tissues and cell lines. So while the breadth of samples is exceptional, the number of clones per miRNA is not sufficient for determining the frequency of RNA editing. For example, for miR-363 there were three edited clones out of 133 clones across the panel of 143 libraries, which is less than one clone of miR-363 per library. In addition, two of the edited clones occurred in one tissue sample which had only six clones for miR-363 which gave a frequency of 33% editing in this sample. This low number of clones per tissue is not unique to miR-363 but is rather a general feature of the cloning libraries. This raises the question of how reliable a low clone number is for ascertaining editing frequencies. Overall, occurrence of edited mature miRNA ranged from 38 out of 8837 clones to 19 out of 24 clones for an individual miRNA across all samples. In some cases the editing was 100 percent in a particular tissue or as low as one clone in 230 in another tissue.

Other work has investigated brain-specific samples,[17] bovine miRNA[18] and T-cells.[19] The work by Kawahara and colleagues examined pri-miRNA in human brain and found that 16% of pri-miRNA transcripts were edited. In contrast, bovine mature miRNAs derived from different tissues showed less editing, as did clones of mature miRNAs from T-cells. Explanations for the discrepancies could be that editing of pri-miRNAs does not lead to production of edited mature miRNAs, or that pri-miRNAs are simply longer with more positions that can be edited in comparison to mature miRNAs.

To conclude: with the rise in high-throughput sequencing projects it has become apparent that there is sequence variation in miRNAs—however the source this variation is unknown in most cases. A-to-I editing, by ADARs, does account for some of this variation. A database detailing the variation of miRNA sequences would clarify the actual abundance of A-to-I edited miRNA. However the overall level of RNA editing is not as important as the biological consequences of miRNA editing. Editing of the Q/R site in the *GluR-B* transcript is a prime example of biological importance. Even though there are not many transcripts edited by ADAR2 in mammals the consequence of editing this one transcript changes the properties not only of the encoded protein but of AMPA receptors, the workhorse class of excitatory glutamate receptors, throughout the CNS. So rather than focus on prevalence a better issue to address is whether there is an equivalent to *GluR-B* editing amongst the 10% of miRNAs that are edited?

EFFECTS OF EDITING OF pri-miRNAs AND pre-miRNAs ON BIOGENESIS

For the majority of edited pri-miRNAs and pre-miRNAs, the consistent outcome has been a decrease in the levels of mature miRNA. Inosine appears to be a negative modulator of miRNA production both at the Drosha microprocessor level, where pri-miRNA transcripts are cleaved to produce pre-miRNAs, and at the Dicer-TRBP level, where the pre-miRNA hairpin is cleaved to produce mature duplex miRNA (Fig. 2).

The first miRNA shown to be edited was miR-22. In a screen of EST clones, Maas and colleagues found editing of pri-miR-22[20] and confirmed that this occurred in vivo. This suggested that A-to-I editing of miRNA can take place prior to Drosha-mediated processing and supported the hypothesis that A-to-I editing could alter Drosha-mediated processing. This hypothesis was proven when it was found that in an in vitro cleavage assay the presence of an edited nucleotide inhibited the ability of the Drosha-DGCR8 complex to process pri-miR-142.[21] Even with as little as 7% of the pri-miR-142 transcript

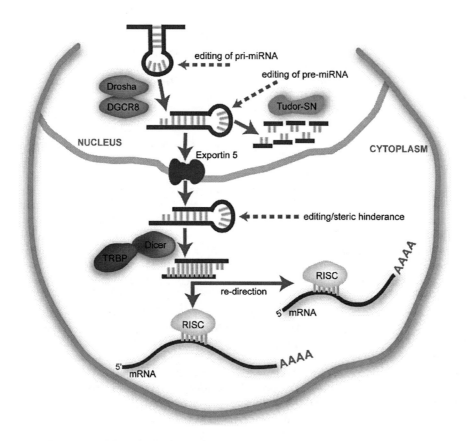

Figure 2. Nuclear editing of pri-miRNAs by ADARs can lead to the disruption of Drosha-DGCR8 processing. Also, editing of pre-miRNAs in the nucleus may lead to export of edited pre-miRNAs or degradation by Tudor-SN. In the cytoplasm, cytoplasmic ADARs, like ADAR1 p150, can potentially edit pre-miRNAs. So far, editing of pre-miRNAs has been observed to lead to disruption of Dicer-TRBP processing and subsequent reduction in mature miRNA levels. Finally, if edited mature miRNA is made, editing can lead to miRNA redirection to alternative target mRNAs.

edited, there was a marked decrease in production of pre-miR-142. In further experiments the authors transfected HEK293 cells using a construct expressing pri-miR-142 in which guanosines replaced the edited adenosines. 'Pre-editing' by mutating the edited A to G led to the accumulation of the mutated pri-miR in transfected HEK293 cells and a concomitant loss of mature miRNA. Thus, the consequence of editing pri-miR-142 was inhibition of cleavage by Drosha-DGCR8.

Conversely, a recent study examining miRNA edited in brain found that pri-miR-203 editing actually enhanced the in vitro cleavage by the Drosha-DGCR8 complex.[17] This is the only example to date of enhanced miRNA production through A-to-I editing. Also, in this study the authors found additional miRNA editing events which inhibited Drosha-DGCR8 processing and the authors noted that evidence of in vivo expression of mature miR-203 was difficult to obtain. The final caveat is that the cleavage reactions were done in vitro with pre-edited pri-miRNA in the absence of ADARs, whose binding alone can influence cleavage activity.[22]

Analysis of brain specific miRNAs revealed also examples of Dicer-TRBP complex inhibition by RNA editing. Processing of pre-let-7g by Dicer-TRBP was shown to be decreased by replacing the adenosine at the editing site with guanosine.[17] A similar example is pre-miR-151 where processing was also impaired by A-to-I editing.[23] In this study, synthetic RNA containing inosine at the edited position was tested for cleavage by Dicer-TRBP in vitro. While Dicer-TRBP could still bind edited pre-miR-151, the presence of inosine at the +3 and/or -1 site restricted the production of mature miR-151. This provides an explanation for the accumulation of edited pre-miR in human Amygdala (38% editing in pri-miRNA, 100% in pre-miRNA) and in mouse cerebral cortex (13% editing in pri-miRNA, 94% editing in pre-miRNA).

Finally, Tudor-SN, a staphylococcal nuclease which has been copurified with RISC,[24] can cleave inosine-containing RNA duplexes.[25] Intriguingly, inhibition of Tudor-SN by 2'-deoxythimidine-3',5'-bisphosphate (pdTp) results in the accumulation of edited pri-miR-142 in HEK293 cells that overexpress pri-miR-142.[21] Therefore A-to-I editing, in addition to inhibiting cleavage by the Drosha microprocessor or Dicer/TRBP, may also generate a substrate for degradation of miRNAs by Tudor-SN.

Although A-to-I editing appears to have a largely inhibitory role on mature miRNA production, edited mature miRNA are generated, as they have been cloned. For example, Pfeffer et al. reported that 12 of 14 mature KSHV virus miR-K12-10 clones had an editing event.[26] Also as mentioned previously, edited mature miRNAs have been observed by Kawahara[27] and by Landgraf.[16] Thus, although editing is primarily inhibitory to processing, A-to-I editing can result in the production of mature miRNAs with altered sequences.

THE EFFECT OF EDITING ON miRNA FUNCTION

For RNA editing effects on miRNA biogenesis or targeting to be significant, they must produce detectable alterations in the final miRNA activity. Even though miR-22 was one of the first miRNAs found to be edited and even though it is widely expressed in mammalian tissues, the consequences of this editing event have not been explored.[20] One miRNA that has been studied is pri-miR-142, which is expressed in the T-lymphoid lineage and edited by both ADAR1 and ADAR2 at 11 different sites.[21] Drosha cleavage is impaired and the edited pri-miRNA is degraded by Tudor-SN. The level of mature miR-142 was seen to substantially increase in both *ADAR1* and *ADAR2* null mice. However there is no obvious corollary effect on the T-lymphoid lineage in these mutant mice.

This exemplifies the problem that despite the evidence that miRNAs are edited, when one critically analyses the functional data, there is a lack of examples where editing has been demonstrated to actually affect the downstream miRNA activity in vivo. Even where it is demonstrated that RNA editing can affect biogenesis of a particular miRNA, there is probably redundancy within the miRNA network. Thus, other miRNAs could affect the same target so the reduction in the amount of one miRNA may not have a pronounced effect.

The most direct effect by an edited miRNA on downstream miRNA activity was reported by Kawahara and colleagues.[27] They demonstrated that editing within the seed sequence in the mature miRNA-376a-5p alters the specificity of the miRNA (Fig. 1) and this can redirect its silencing activity to a new set of targets transcripts such as the transcript encoding phosphoribosyl pyrophosphate synthetase 1, which is involved in uric acid synthesis. The authors elegantly showed that in *ADAR2* null mice there is a two-fold increase in uric acid levels in the cortex. .

ADARs AS COMPETING dsRNA-BINDING PROTEINS

Heale and colleagues analysing the editing of pri-miRNA-376a2 demonstrated that it was not editing of this pri-miRNA that interfered with Drosha cleavage activity in vitro (Fig. 1) but the binding of ADAR2 to the pri-miRNA.[22] These experiments suggested that ADARs could have a greater influence on miRNA biogenesis than previously anticipated and that they could affect more than the 10% miRNAs that are found to be edited. It also reiterated what had been previously observed that binding of proteins with dsRBDs to dsRNA is rather nonspecific so that competition of dsRBDs for dsRNA has significant effects. In fact, RNA-binding proteins can strongly regulate miRNA biogenesis as has been observed for hnRNP A1 (see chapter by Michlewski et al).[28]

This study also raised the possibility that other dsRBD-containing proteins may have a role in regulating miRNA biogenesis. ADAR3 and TENR do not have enzymatic activity yet they have dsRBDs and are evolutionarily conserved. In addition there are other dsRBD proteins in various genomes that as yet have no assigned function. However, it is possible that there is redundancy between these proteins so it may require multiple knockouts to reveal such functions.

EDITING OF SEED TARGET SEQUENCES WITHIN 3'UTRs

The effect of editing might be greater when the miRNA target sequence within the 3'UTR is edited rather than the miRNA itself. It has long been noted that 90% of mammalian RNA editing occurs within 3'UTR regions of protein-coding transcripts. Heale and colleagues using a dual luciferase reporter assay found that when the target sequence of a miRNA is replaced with a sequence corresponding to an edited version of the miRNA this had a greater effect on activity than editing the miRNA itself.[22]

Recently 12,723 editing sites within human 3'UTRs were screened to investigate if editing either created or destroyed miRNA seed recognition sequences.[29] Over 3,000 editing events that were found within 3′UTRs were indeed within potential seed recognition sequences. Particularly in the case of 200 ESTs, editing of a specific sequence motif created a seed match to three unrelated miRNAs. This suggests that one outcome of editing 3′UTRs is to modulate miRNA target recognition. What is very intriguing is that in *C. elegans* no editing has been identified that modifies codons within transcripts—however there is extensive editing in both 3′and 5′UTRs of protein-coding transcripts.[30] This raises the possibility that editing can modulate miRNA target sequences in *C. elegans*.

CONCLUSION

There is strong evidence that editing can affect the biogenesis of miRNAs at different stages. Nuclear editing of pri-miRNAs by ADARs can lead to the disruption of Drosha-DGCR8 processing. At the next step, editing of pre-miRNAs in the nucleus may lead to degradation by Tudor-SN or to export of edited pre-miRNAs. In the cytoplasm, cytoplasmic ADARs, like ADAR1 p150, can also potentially edit pre-miRNAs. So far, editing of pre-miRNAs has been observed to lead to disruption of Dicer-TRBP-mediated processing and subsequent reduction in mature miRNA levels. Finally, if edited mature miRNA is made, editing can lead to miRNA redirection to alternative target mRNAs.

In general, editing appears to reduce the amount of mature miRNA, yet despite this there is only one example where it has been demonstrated unequivocally that editing affects the downstream function of the miRNA. It is likely there are more examples as we are still in the early days of analysing the function of miRNAs.

It is possible that editing miRNA target sequences within the 3′UTRs may have greater consequences than editing individual miRNAs. Also ADAR RNA editing enzymes are first and foremost dsRNA-binding proteins and this may ultimately be as important as the effect of their enzymatic activity in miRNA biogenesis.

ACKNOWLEDGEMENTS

Thanks to Craig Nicol for assistance with figures. MO'C is supported by core funding from the Medical Research Council (U.1275.01.005.00001.01). BH is supported by a Fellowship from the Marie Curie Foundation (PIIF-GA-2008-220317).

REFERENCES

1. Heale BSE, M.A.O'Connell. Biological Roles of ADARs. In: Grosjean H, ed. DNA and RNA Modification Enzymes: Structure, Mechanism, Function and Evolution. Austin: Landes Bioscience; 2009; 243-258.
2. Sommer B, Kohler M, Sprengel R et al. RNA editing in brain controls a determinant of ion flow in glutamate-gated channels. Cell 1991; 67(1):11-19.
3. Levanon EY, Eisenberg E, Yelin R et al. Systematic identification of abundant A-to-I editing sites in the human transcriptome. Nat Biotechnol 2004; 22(8):1001-1005.
4. Adams MD, Celniker SE, Holt RA et al. The genome sequence of Drosophila melanogaster. Science 2000; 287(5461):2185-2195.
5. Venter JC, Adams MD, Myers EW et al. The sequence of the human genome. Science 2001; 291(5507):1304-1351.

6. Patterson JB, Samuel CE. Expression and regulation by interferon of a double-stranded-RNA-specific adenosine deaminase from human cells: evidence for two forms of the deaminase. Mol Cell Biol 1995; 15(10):5376-5388.
7. Steitz TA. Similarities and differences between RNA and DNA recognition by proteins. In: Gesteland RF, Atkins JF, eds. "The RNA World". Cold Spring Harbor: Cold Spring Harbor Laboratory Press; 1993:219-237.
8. Bevilacqua PC, Cech TR. Minor-groove recognition of double-stranded RNA by the double-stranded RNA-binding domain from the RNA-activated protein kinase PKR. Biochemistry 1996; 35(31):9983-9994.
9. Reid JG, Nagaraja AK, Lynn FC et al. Mouse let-7 miRNA populations exhibit RNA editing that is constrained in the 5′-seed/cleavage/anchor regions and stabilize predicted mmu-let-7a:mRNA duplexes. Genome Res 2008; 18(10):1571-1581.
10. Lu S, Sun YH, Chiang VL. Adenylation of plant miRNAs. Nucleic Acids Res 2009; 37(6):1878-1885.
11. Li J, Yang Z, Yu B et al. Methylation protects miRNAs and siRNAs from a 3′-end uridylation activity in Arabidopsis. Curr Biol 2005; 15(16):1501-1507.
12. Ramachandran V, Chen X. Degradation of microRNAs by a family of exoribonucleases in Arabidopsis. Science 2008; 321(5895):1490-1492.
13. Heo I, Joo C, Cho J et al. Lin28 mediates the terminal uridylation of let-7 precursor microRNA. Mol Cell 2008; 32(2):276-284.
14. Katoh T, Sakaguchi Y, Miyauchi K et al. Selective stabilization of mammalian microRNAs by 3′ adenylation mediated by the cytoplasmic poly(A) polymerase GLD-2. Genes Dev 2009; 23(4):433-438.
15. Blow MJ, Grocock RJ, van Dongen S et al. RNA editing of human microRNAs. Genome Biol 2006; 7(4):R27.
16. Landgraf P, Rusu M, Sheridan R et al. A mammalian microRNA expression atlas based on small RNA library sequencing. Cell 2007; 129(7):1401-1414.
17. Kawahara Y, Megraw M, Kreider E et al. Frequency and fate of microRNA editing in human brain. Nucleic Acids Res 2008; 36(16):5270-5280.
18. Jin W, Grant JR, Stothard P et al. Characterization of bovine miRNAs by sequencing and bioinformatics analysis. BMC Mol Biol 2009; 10:90.
19. Wu H, Neilson JR, Kumar P et al. miRNA profiling of naive, effector and memory CD8 T-cells. PLoS One 2007; 2(10):e1020.
20. Luciano DJ, Mirsky H, Vendetti NJ et al. RNA editing of a miRNA precursor. RNA 2004; 10(8):1174-1177.
21. Yang W, Chendrimada TP, Wang Q et al. Modulation of microRNA processing and expression through RNA editing by ADAR deaminases. Nat Struct Mol Biol 2006; 13(1):13-21.
22. Heale BS, Keegan LP, McGurk L et al. Editing independent effects of ADARs on the miRNA/siRNA pathways. EMBO J 2009; 28(20):3145-3156.
23. Kawahara Y, Zinshteyn B, Chendrimada TP et al. RNA editing of the microRNA-151 precursor blocks cleavage by the Dicer-TRBP complex. EMBO Reports 2007; 8(8):763-769.
24. Caudy AA, Ketting RF, Hammond SM et al. A micrococcal nuclease homologue in RNAi effector complexes. Nature 2003; 425(6956):411-414.
25. Scadden AD. The RISC subunit Tudor-SN binds to hyper-edited double-stranded RNA and promotes its cleavage. Nat Struct Mol Biol 2005; 12(6):489-496.
26. Pfeffer S, Sewer A, Lagos-Quintana M et al. Identification of microRNAs of the herpesvirus family. Nat Methods 2005; 2(4):269-276.
27. Kawahara Y, Zinshteyn B, Sethupathy P et al. Redirection of silencing targets by adenosine-to-inosine editing of miRNAs. Science 2007; 315(5815):1137-1140.
28. Guil S, Caceres JF. The multifunctional RNA-binding protein hnRNP A1 is required for processing of miR-18a. Nat Struct Mol Biol 2007; 14(7):591-596.
29. Borchert GM, Gilmore BL, Spengler RM et al. Adenosine deamination in human transcripts generates novel microRNA binding sites. Hum Mol Genet 2009; 18(24):4801-4807.
30. Morse DP, Aruscavage PJ, Bass BL. RNA hairpins in noncoding regions of human brain and Caenorhabditis elegans mRNA are edited by adenosine deaminases that act on RNA. Proc Natl Acad Sci USA 2002; 99(12):7906-7911.

CHAPTER 9

miRNAs NEED A TRIM
Regulation of miRNA Activity by
Trim-NHL Proteins

F. Gregory Wulczyn,* Elisa Cuevas, Eleonora Franzoni
and Agnieszka Rybak

Abstract: Trim-NHL proteins are defined by RING, B-Box and Coiled-coil protein motifs
(referred to collectively as the Trim domain) coupled to an NHL domain. The *C.
elegans, D. melanogaster*, mouse and human Trim-NHL proteins are potential
and in several cases confirmed, E3 ubiquitin ligases. Current research is focused
on identifying targets and pathways for Trim-NHL-mediated ubiquitination and
in assessing the contribution of the NHL protein-protein interaction domain
for function and specificity. Several Trim-NHL proteins were discovered in
screens for developmental genes in model organisms; mutations in one of the
family members, Trim32, cause developmental disturbances in humans. In most
instances, mutations that alter protein function map to the NHL domain. The NHL
domain is a scaffold for the assembly of a translational repressor complex by the
Brat proto-oncogene, a well-studied family member in *Drosophila*. The link to
translational control is common to at least four Trim-NHLs that associate with
miRNA pathway proteins. So far, two have been shown to repress (Mei-P26 and
Lin41) and two to promote (NHL-2, Trim32) miRNA-mediated gene silencing. In
this chapter we will describe structure-function relations for each of the proteins
and then focus on the lessons being learned from these proteins about miRNA
functions in development and in stem cell biology.

*Corresponding Author: F. Gregory Wulczyn—Center for Anatomy, Institute of Cell Biology
and Neurobiology, Charité—Universitätsmedizin Berlin, Berlin, Germany.
Email: gregory.wulczyn@charite.de

Regulation of microRNAs, edited by Helge Großhans.
©2010 Landes Bioscience and Springer Science+Business Media.

INTRODUCTION

The year 2000 was an auspicious one for the miRNA field. In a series of three papers, the let-7 miRNA was first identified in a screen for mutations affecting developmental timing in *C. elegans*.[1] Second, *let-7* was quickly demonstrated to be highly conserved and expressed in a wide range of bilaterally symmetric animals,[2] unlike the original miRNA gene, *lin-4*, that was thought to be restricted to *C. elegans*.[3] Third, *let-7* was shown to act as a repressor for a conserved gene, *lin-41*,[1,4] within the developmental timing pathway. Analysis of predicted protein domains in LIN-41 led to the description of a novel class of so-called Ring, B-Box, Coiled-coil (RBCC) domain proteins.[4,5] Later, the acronym Trim (*Tri*partite *m*otif) was adopted for this sequence of protein domains. Current annotations recognize over 70 Trim domain proteins in mammalian genomes that can be subdivided into distinct classes based on the presence of additional C-terminal domains (reviewed in refs. 6,7). The C-terminus of LIN-41 was found to contain six copies of a 44 amino acid repeat sequence shared with two other known proteins: HT2A and NCL-1.[4,5] These were the founding members of the Trim-NHL (*N*CL-1, *H*T2A, *L*IN-41) family. As seen in Figure 1, there are four Trim-NHL proteins encoded in the *D. melanogaster*, *M. musculus* and *H. sapiens* genomes and five in *C. elegans*. Several of these proteins,

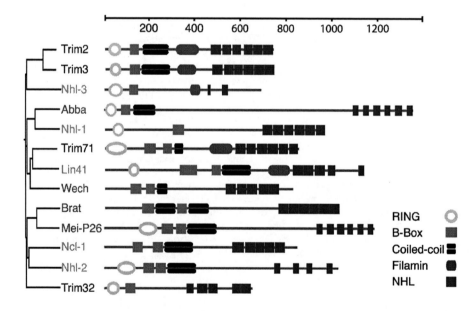

Figure 1. Overview of the Trim-NHL protein family. The mammalian NHLrc1-3 proteins are not considered due to space constraints and because of the lack of Trim motifs in these proteins (with the exception of the RING in NHLrc1). Protein motifs were identified using the SMART (http://smart.embl-heidelberg.de), InterProScan (http://www.ebi.ac.uk/Tools/InterProScan/) and ProSitescan (http://expasy.org/tools/scanprosite/) tools. A phylogenetic tree was generated after ClustalW alignment of the C-terminal sequences. The relationships may vary from published phylogenies depending on the sequences chosen for alignment and precise methodologies used. For display, the *x*-axis was compressed six-fold. *C. elegans* proteins are in light gray (green), *D. melanogaster* in medium gray (red), *M. musculus* in black (blue) type. A color version of this figure is available at www.landesbioscience.com/curie.

including the mouse ortholog of LIN-41, have recently been identified as regulators of miRNA pathway activity, the subject of this review.

There are several reviews of Trim domain proteins, including their structural characterization, disease phenotypes and functions in ubiquitin-mediated protein degradation.[6-8] As of this writing, however, there is no comprehensive review of Trim-NHL family members and their roles in development and miRNA regulation. For this reason, we will begin with a brief outline of the common structural features of this protein family, followed by a review of earlier work on the functional characterization of each of the members. We will then turn to recent evidence that at least some of the Trim-NHL proteins act as post-transcriptional regulators of miRNA activity in their capacity as E3 ubiquitin ligases.

THE Trim-NHL FAMILY OF DEVELOPMENTAL REGULATORS

Although there have been a number of important studies on Trim-NHL proteins in chicken, zebrafish or rat, the bulk of the work has concerned the proteins from *C. elegans*, *Drosophila*, mouse and humans presented in Figure 1. The phylogenetic tree presented in Figure 1 is based on alignment of the C-terminal NHL sequences. Both the phylogenetic relationships and the individual domain structures of these proteins have been more rigorously treated elsewhere.[6,7,9-11] As detailed below, the Trim-NHL sequence of motifs was first recognized in LIN-41 from *C. elegans* and consists of a RING-type (*R*eally *I*nteresting *N*ew *G*ene) Zinc-finger in close proximity to the N-terminus. One or two copies of a distinct Zinc-finger motif referred to as the B-Box follow the RING. The triad is completed by a hydrophobic heptad repeat termed the coiled-coil. This linear arrangement is quite constant, suggesting cooperative functional interactions between the three motifs. Less conserved is the Ig-filamin domain situated between the Trim motif and the NHL domain in several of the proteins. The paradigm NHL domain contains six copies of a 44 residue repetitive motif but the number of readily identified repeats varies from 2 to 6.

Each of these motifs will be discussed in turn, beginning with the RING domain. Although referred to as a RING finger, the actual structure is not fingerlike. Instead, the RING sequence contains four pairs of Zinc binding residues (numbered 1-4). The peptide chain loops back upon itself to juxtapose two non-adjacent pairs of Zinc binding residues (pairs 1 and 3) to form the first Zinc-coordination site. The peptide chain then reverses direction again to align the two remaining pairs (pairs 2 and 4). The result is a rigid, self-reinforced globular structure referred to as a "cross-brace" (reviewed in refs. 12,13; see NCBI Conserved Domain Database 00162). RING motifs are found in a large number of proteins and were originally thought to mediate DNA binding and/or protein-protein interactions. After several RING proteins were linked to ubiquitin-mediated protein degradation, the RING domain of the Cbl (Casitas B-lineage lymphoma) proto-oncogene was shown to actively participate in protein ubiquitination,[14,15] a finding then extended to many additional family members[16] (reviewed in refs. 13,17). The RING domain was later shown to possess intrinsic E3 ubiquitin ligase activity and to directly recruit and activate specific E2 ubiquitin conjugating enzymes.[18,19] A brief description of the ubiquitin pathway will be provided in the next section (see Fig. 2). However, it seems certain that E3 activity is a general feature of the RING domain. Trim32 was the first Trim-NHL

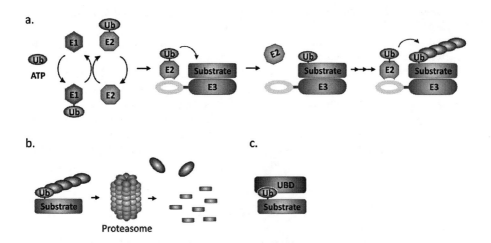

Figure 2. Basic principles of ubiquitination. a) A ubiquitin monomer is covalently linked to an E1 activating enzyme. Interaction of the E1 with an E2 conjugating enzyme results in transfer of the primed ubiquitin to an E2 conjugating enzyme. For RING domain E3 ubiquitin ligases, the E2 interacts directly with the RING. In general, the E3 attracts substrates via additional interaction surfaces such as the coiled-coil or NHL domains. Construction of polyubiquitin chains requires iterative cycles of E2 binding and ubiquitin ligation. Ubiquitin is covalently attached via an isopeptide bond between the C-terminal glycine of ubiquitin and a substrate lysine. See text and references for details. b) Current research is revealing new forms of ubiquitin linkage. In the classical pathway depicted here linear chains of at least four residues are connected at Lys48 of ubiquitin. This linkage is specifically recognized by the 26S proteasome, resulting in ubiquitin release for recycling and proteolytic degradation of the tagged substrate. c) One alternative to the classical pathway is schematically shown. Monoubiquitination can support protein binding with a partner containing a ubiquitin binding domain (UBD), allowing dynamic regulation analogous to protein phosphorylation.

protein shown to be an E3 ubiquitin ligase,[20,21] followed by Trim2[22] and mouse LIN-41 (referred to as either Trim71, mLin41, or Mlin41).[23]

Less is known about the B-Box motifs, which are found either singly or in tandem. When in tandem, the upstream B-Box conforms to a distinct consensus sequence. B-Boxes are most commonly associated with a RING domain, but as seen in Brat and Dappled/ Wech, this is not always the case (Fig. 1). Mutations in the B-Box for the non-NHL Trim proteins Trim5a and Trim18 (Midline-1) disrupt protein function (reviewed in ref. 6). As discussed below, mutations in the B-Box of human Trim32 lead to a distinct developmental disorder (Bardet-Biedl Syndrome)[24] compared to mutations in the NHL domain (Limb Girdle Muscular Dystrophy).[21,25] However, in these and in other cases the precise role of the B-Box remains elusive and may entail protein turnover, oligomerization or intracellular localization in addition to specific protein-protein interactions. The solution structure of the individual and the tandem B-Boxes of Trim18 is an important advance, revealing that both adopt a cross-brace conformation quite similar to the RING domain[26-28] (see also NCBI Conserved Domain Database cl00034). This raises the interesting possibility that the B-Box may directly interact with the RING and participate in E3 activity, perhaps by serving as an accessory binding surface for E2 conjugating enzymes.

Of the three Trim motifs, the coiled-coil is the most widespread outside the Trim superfamily and represents a basic building block of protein structure (reviewed in refs. 18,29). In principle, coiled-coils can form higher order homo- and heteromeric structures with a high degree of plasticity.[29] Less is known about their structural or functional contribution to Trim domain function, although there is experimental evidence that they support homomeric interactions.[9] The same study showed that multimerization mediated by the coiled-coil was required for correct localization to intracellular compartments. Trim2, Trim3 and Trim32 eGFP fusions were found to be cytoplasmic with a filamentous, diffuse or "speckled" distribution, respectively.[9]

When present, the Filamin homology domain, also referred to as an Ig-filamin repeat (see NCBI Conserved Domain Database cl02665), lies between the Trim motifs and the NHL repeats. No function has been assigned to the Filamin domain in the context of Trim-NHL family members. In Filamin proteins, the Ig-Filamin repeats are present in up to twenty-four copies and adopt a rod-like β-barrel conformation that interacts with actin and many other partners (see ref. 30 for a recent review).

Of course, the defining domain of the Trim-NHL family is the NHL repeat (Fig. 1). The similarity of the NHL repeats to a previously characterized motif, the WD-40 β-propeller, was significant enough to suggest a similar secondary structure.[5] Furthermore, many of the genetically isolated mutations in LIN-41 and Brat mapped to the NHL domain.[4,31] The NHL domain was necessary and sufficient for rescue of *brat* embryos during embryonic patterning,[32] demonstrating that the isolated domain has activity as a translational repressor (see below for details). The crystal structure of the isolated NHL from Brat has been solved and revealed that each of the NHL repeats forms one "blade" of a six-bladed propeller, with the exception of the first blade. β-sheet elements derived from the first and sixth NHL repeat contribute to the first blade. This brings both ends of the structure together in a circular, or doughnut-like arrangement.[33] Each blade is composed of four β-sheets connected by exposed loops. Structural modeling suggests that the RNA-binding protein Pumilio interacts with one face of the propeller via the looped out residues. The properties of this potential interaction surface are predicted to vary substantially among the individual Trim-NHL family members such as Dappled/Wech.[33]

THE Trim DOMAIN AS E3 UBIQUITIN LIGASE

As noted above, E3 ubiquitin ligase activity has been documented for an increasing number of Trim proteins, making it the potentially largest class of E3 enzymes. In the case of Trim-NHL proteins, a role in protein ubiquitination could account for the diversity of their known functions in processes ranging from developmental timing (LIN-41, NHL-2), patterning (Brat), cell growth, division and proliferation (Brat, Mei-P26, NCL-1), endosomal trafficking (Trim3), muscle integrity (Trim32, Wech) and germ line (LIN-41, Mei-P26) or nervous system development and function (Brat, Mei-P26, mLin41, Trim2, Trim32). Ubiquitination is a post-translational modification rivaling phosphorylation in extent and consequence. The ubiquitin literature is far too vast to summarize here and there are many excellent reviews.[13,34,35] Basic features relevant to this review are summarized in Figure 2. Covalent attachment of ubiquitin to substrate proteins involves a tag team of three enzymatic activities (E1-E3). First, in an ATP-consuming reaction ubiquitin is activated by attachment to an E1 activating enzyme. Interaction with an E2 conjugating enzyme leads to transfer of ubiquitin from the E1 to the active site of the E2. E3 ligases belong to one of two large classes, HECT or RING. In the case of RING domain E3s, the E3 serves as a bridge between the E2 and the substrate protein with primary responsibility for substrate specificity. Trim E3s are believed to stimulate catalysis by inducing proximity between the E2 and the substrate.[13] A total of six known E1s (only two utilize ubiquitin, the others employ ubiquitin-like proteins such as SUMO—*S*mall *U*biquitin-like *MO*difier) can chose among ~40 E2s. A given E3 will productively interact with only one or a few E2s and can recognize multiple substrate proteins. Substrate ubiquitination can take many forms, including monoubiquitination, oligoubiquitination and various types of polyubiquitin chains (Fig. 2). To date, for the Trim-NHLs there is experimental evidence for polyubiquitination activity using Lys48 linkages, one of seven lysines available in ubiquitin for chain elongation.[21-23,36,37] This is the classical signal for recognition and degradation by the 26S proteasome. Alternative linkages are nonproteolytic and alter substrate function, for example in directing protein traffic to vesicles (monoubiquitination) or by acting as a scaffold for protein-protein interactions (Lys63 polyubiquitination). Since linkage specificity is primarily a function of the E2, an E3 can have dual capacity and there is experimental support for substrate monoubiquitination via Trim32.[38]

FUNCTIONAL ANALYSIS OF INDIVIDUAL Trim-NHL FAMILY MEMBERS

Trim32

Trim32 was the first Trim-NHL gene to be cloned, in a two-hybrid screen for proteins interacting with the HIV Tat protein.[39] A missense mutation in the Trim32 NHL domain was later shown to underlie the human developmental disturbance Limb-girdle muscular dystrophy (LGMD2H, Type 2H).[40] Subsequent studies have identified additional Trim32 mutations in LGMD2H; all map to the NHL and affect both the ability of the protein to self-interact and to bind E2.[25] In keeping with the muscle phenotype, Trim32 was originally shown to physically interact with the head and neck region of the myosin heavy chain. The coiled-coil domain was required for myosin binding. Trim32 was shown to possess E3 activity in an in vitro autoubiquitination assay, in conjunction with the E2 enzymes UbcH5a, UbcH5c or UbcH6. Based on in vitro and in vivo assays, a role for Trim32 in actin degradation was proposed during muscle regeneration.[21] More recently, Dysbindin was identified as an additional candidate for Trim32-mediated ubiquitination, primarily in the form of monoubiquitination.[38] Dysbindin is a component of Bloc-1, an actin-associated protein complex involved in the biogenesis of lysosome-related vesicles.[41]

An alternative model was proposed for Trim32 in keratinocytes, in which the NHL domain of Trim32 interacts with Piasy. Piasy is itself an E3 ligase, but is specific for SUMO,[36] an alternative ubiquitin-like protein tag with a distinct spectrum of activities. Interestingly, Trim32 not only targeted Piasy for ubiquitin-mediated degradation, it also caused the relocalization of Piasy from the nucleus to cytoplasmic foci resembling P-bodies. A consequence of Trim32 expression in keratinocytes was reduced sensitivity to NF-κB dependent apoptosis, suggesting a role for the Trim32/Piasy regulatory interaction in skin carcinogenesis.[36]

It is not surprising that an E3 ligase might have multiple substrates, with the possibility of multiple gene-phenotype relationships. This point is underscored by an independent mutation in the B-Box domain of Trim32 that predisposes for the Bardet-Biedl Syndrome.[24] Bardet-Biedl is a multifaceted disorder encompassing renal abnormalities, retinal dystrophy, polydactyly, mental retardation and obesity. Nevertheless, there is no obvious phenotypic overlap between LGMD2H and Bardet-Biedl and none of the many Bardet-Biedl genes other than Trim32 have any known connection to muscular dystrophy.

However, the finding that Bardet-Biedl genes as a group cause defects in the basal body of ciliated cells and impair ciliary function[42,43] may be relevant to the proposed role for Trim32 in neurogenesis,[37] which is discussed below. New insights are likely to be gained by transgenic approaches, for example deletion of Trim32 in mice was recently shown to phenocopy many of the myopathies associated with LGMD2H.[44] The phenotype also has a neurogenic component, manifested biochemically by reduced levels of neurofilaments and morphologically by reduced diameters of motor axons. Bardet-Biedl symptoms were not described. One caveat for comparing the human mutations with the gene deletion model is that the mutations in Bardet-Biedl or in LGMD2H are not necessarily null alleles. Therefore, it may be necessary to generate mice carrying NHL or B-Box mutations to unravel the full functions of Trim32.

LIN-41

In *C. elegans*, *lin-41* was discovered as a major downstream target for the *let-7* miRNA in the developmental timing pathway[1,4] (reviewed in ref. 45). Gain and loss of function *lin-41* mutants displayed opposite phenotypes affecting vulval development and morphology, oocyte production and the timing of cell cycle exit and terminal differentiation of a set of blast cells in the hypodermis known as seam cells.[4] Genetic analysis gave the first indication that LIN-41 is itself a post-transcriptional regulator, acting to repress a transcription factor downstream of *lin-41* in the timing pathway. Sequencing of *lin-41* led to the recognition of the Trim-NHL domain structure as a new protein family, together with the previously cloned NCL-1 and HT2A/Trim32.[4,5] Many of the null mutations recovered in the genetic screen mapped to the NHL domain, a clear indication of its functional importance. Ultimately, the presence of the Trim domain led to the annotation of mammalian homologs of LIN-41 as Trim71, but here we retain the designation Lin41 for the mouse and human proteins in deference to this pioneering work in *C. elegans*.

Database searches revealed the presence of LIN-41 homologs in *Drosophila* and in vertebrates.[4] In zebrafish, Kloosterman et al. verified the prediction by Slack and coworkers that regulation of *lin-41* by *let-7* is conserved.[46] Three papers described the cloning of the chicken, mouse and human genes and analyzed embryonic mRNA expression patterns.[47-49] In both mouse and chicken, *Lin41* was temporally regulated and inversely correlated with the induction of let-7 and miR-125 expression in embryonic development.[48,49] In *C. elegans*, *lin-41* is expressed in neurons, muscles and gonads.[4] Muscle expression was confirmed in the chicken[47] and all three research groups noted temporally dynamic regulation of the mRNA in the limb, wing and tail buds.[47-49] Consistent with this, *mLin41* expression was modified in mouse mutants for the limb bud patterning genes Sonic hedgehog, Fgf-8 and Gli3.[47] This represents the first evidence for developmental control of *Lin41* and its miRNA regulators let-7 and miR-125 by mammalian signaling pathways.

In *Drosophila*, the closest *lin-41* homolog has been referred to as either *dappled* or, more recently, *wech*. As seen in Figure 1, the protein lacks a RING domain. Dappled refers to the appearance of mutant larvae, due to the presence of melanotic tumors that are thought to be a sign of tissue abnormalities.[50] *dappled* was next encountered in a differential screen for genes expressed in the embryonic head. At Stage 11, the mRNA was expressed at low levels throughout the embryo and more strongly in neuroblasts, ganglion mother cells and neurons.[51] A more comprehensive *in situ* study documented expression in the central nervous system (CNS) and peripheral nervous system (PNS) throughout embryogenesis.[11] Ectopic *dappled* expression interfered with proper development of the eye and sensory organs.[11,52] Although the significance of these findings for normal development remains unclear, the eye phenotype was sensitive to *let-7* expression levels.[11]

A quite different view emerged from studies of integrins and muscle attachment[53] (reviewed in refs. 54,55). Analysis of a mutant with a near complete loss of embryonic muscle attachment identified a loss of function mutation in a gene identical to *dappled* that the authors named *wech*. Loss of Wech had no obvious effect on muscle cell differentiation, as myoblasts fused and expressed actin and myosin heavy chain. The attachment defect could be rescued by specifically restoring Wech expression in muscle and tendon cells, or in either cell-type individually. The Wech protein colocalized to cortical foci in muscle cells with the adhesion proteins Talin, ILK (*I*ntegrin *L*inked *K*inase) and Tensin. These three proteins act to link integrin to the actin cytoskeleton at focal adhesion points between cells known as hemiadherens junctions. Wech may serve as an adaptor protein in the organization of these foci, as ILK and Tensin localization was dependent on Wech and Wech in turn was dependent on integrin and Talin. Interestingly, the B-Box and coiled-coil domains were required for the interaction with both ILK and Tensin.

The adhesion junctions studied by Löer et al are by no means restricted to the muscle-tendon interface, but there is not yet any information on Wech function in other tissues. It may be relevant that Ambros and colleagues found that *let-7* is critical for the attainment of an adult neuromusculature.[56] *let-7* was strongly upregulated in the *Drosophila* CNS and PNS during metamorphosis. High-level expression was also observed in adult body wall muscle and in remnants of larval muscle that are initially retained after metamorphosis. Consistent with this expression pattern, *let-7* null mutants were viable without gross defects in external morphology but suffered from severe disturbances in motor function. Overall CNS development appeared to be normal, but a juvenile pattern of muscular innervation persisted that coincided with a failure of the muscular remodeling that normally occurs upon execution of metamorphosis.[56] Determining the contribution of Wech/Dappled to this transformation is an obvious next step and it will be interesting to see if the protein is downregulated in neurons and muscle at the transition. If so, the question would then be whether another Trim-NHL protein adopts the larval function of Wech/Dappled in adults and what the functional consequences of such a switch might be.

Trim2 and Trim3

As seen in Figure 1, the mammalian Trim2 and Trim3 proteins are more closely related to one another than to their *D. melanogaster* or *C. elegans* paralogs. Like Trim32, both are strongly, but by no means exclusively, expressed in the brain. Originally referred to as Berp (*B*rain *e*nriched *R*ing *P*rotein) Trim3 was cloned in a search for RING finger proteins in the brain.[57] Like Trim32, Trim3 was found to interact with myosin. Unlike Trim32, the NHL domain directed binding to the tail region of the unconventional Class V myosins Va and Vb. Class V myosins are primarily involved in transport of macromolecular complexes, organelles and vesicles, including endosomes (reviewed in ref. 58). Consistent with a role in endosomal trafficking, Trim3 was found to copurify with early endosomes and to form a complex with Actinin-4, Hrs and Myosin Vb dubbed CART (*C*ytoskeleton-*A*ssociated *R*ecycling or *T*ransport).[59] Independently, Trim3 was found to copurify with the Lst2 protein.[60] Focusing on Lst2, Mosesson et al. found that Lst2 promotes trafficking of the receptor for epidermal growth factor (EGFR) to early endosomes. Furthermore, the localization of Lst2 to early endosomes was inhibited by monoubiquitination, but it is not clear if Lst2 is a substrate for Trim3.[60] Mono- and oligoubiquitination are recurring signals in the complex pathways that decide receptor fate after endocytosis: either recycling by return to the cell membrane or degradation by sequential sorting to endosomes and then to lysosomes (reviewed in ref. 61). It remains to be seen if Trim3 has E3 activity, if it targets Lst2 and what its putative role in vesicle trafficking may be.

Trim2 was originally designated NARF (Neural Activity-related Ring Finger), because Trim2 mRNA was upregulated after pharmacological stimulation of seizure activity in rats. Like Trim3, the NHL domain of Trim2 interacted with Myosin V.[62] Within the brain, Trim2 expression is highest in the CA1-3 neurons of the hippocampus,[63] a finding confirmed using a gene-trap allele to record Trim2 (Trim2[GT]) promoter activity.[22] Trim2 is dispensable for embryonic development, as mice homozygous for Trim2[GT] appeared normal until 1.5 months. The mice then presented a neurodegenerative phenotype initially manifested by tremor, followed by ataxia and seizures. Morphologically, Trim2[GT] brains displayed increased accumulation of neurofilament and abnormally swollen axons. This axonopathy preceded both neurological symptoms and loss of neurons in the cerebellum and retina. Biochemically, Trim2 was shown to interact with the E2 enzyme UbcH5a in an autoubiquitination assay,[22] one of the E2 utilized by Trim32.[21] However, unlike Trim32 no autoubiquitination activity was observed with Ubc5Hc or UbcH6. The light chain of neurofilament may be one substrate for Trim2[22] and it will be interesting to compare axonal phenotypes and ubiquitin substrates for Trim2 and Trim32.

Brat

The Drosophila protein Brat (Brain tumor) is perhaps the best studied of the Trim-NHL proteins, due to its dual role as an embryonic patterning gene and as a tumor suppressor in the larval brain. brat mutants also have defects in male and female fertility, cuticle formation and occasionally form melanotic tumors,[64] a constellation reminiscent of lin-41 and dappled, respectively. During early embryonic development Brat is required for the establishment of an anterior-posterior gradient in the translation of hunchback (hb) mRNA, a maternal effect gene and Zinc finger transcription factor.[32] In this role as translational regulator, Brat participates in complex formation with the translational regulator Pumilio and the sequence specific RNA-binding protein Nanos. Despite detailed information regarding the structural features of the regulatory complex, less is known about the mechanistic contribution of Brat to translational control.[33] Sonoda and Wharton have suggested that Brat may assemble on multiple mRNAs by engaging in combinatorial interactions with RNA-binding proteins other than Nanos,[32] but this has not yet been confirmed experimentally. This suggestion may, however, prove to be correct in light of the discovery that Brat interacts with the miRISC effector protein Ago1,[65] as discussed in more detail in the section on Mei-P26.

The most striking feature of brat mutants is the growth of larval tumors in the optic centers that can attain ten times the normal size.[64] Frank et al. found that ectopic Brat inhibits RNA synthesis and cell proliferation while increasing cell size in several organ contexts.[66] However, Brat is primarily expressed in the embryonic PNS, the ventral nerve cord and the developing brain.[31] Even within the CNS, the tumor origin was later shown to be confined to a subset of neural progenitors[67] with the characteristics of transit amplifying cells.[68] A recent review of the relevant issues in the larval nervous system is highly recommended[69] (see also ref. 70) and can only be briefly treated here. In the basic scheme, the intrinsic apical-basal polarity of neuroblasts is used as a guide to asymmetrically sort cell fate determinants to daughter cells (Fig. 3a). Different regions of the fly brain

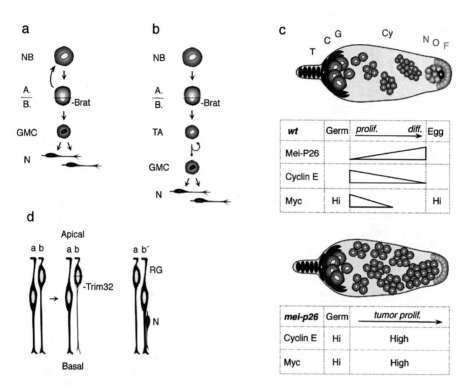

Figure 3. Models for Trim-NHL proteins in stem cell niches. a) In the *Drosophila* CNS, neuroblasts (NB) divide asymmetrically to produce a smaller committed daughter termed ganglion mother cell (GMC) and a new neuroblast. An apical (A.) protein complex (or Par complex) directs segregation of basal (B.) determinants including Brat. In most lineages the GMC divides once to produce two neurons (N). b) In the lineage of origin for Brat tumors (PAN), the initial basal daughter is a transit-amplifying stem cell (TA) able to undergo further division to generate multiple committed GMCs. In the absence of Brat, TA cells fail to differentiate and hyperproliferate. c) A wild-type female *Drosophila* germarium is depicted (*Drosophila* ovarian stem cells are reviewed in ref. 112). Polarity is dictated by the Germ cell niche, comprised of the basement membrane and terminal filament (T), cap cells (C) and germ cells (G). Germ cell daughters, or cytoblasts (Cy), undergo four rounds of transit-amplifying divisions. Cystoblasts remain connected by cytoplasmic bridges (not shown) and become progressively smaller. One cytoblast is determined and differentiates to an oocyte (O, dark). The remaining cystoblasts become nurse cells (N), enclosed in follicular epithelial cells to form an egg chamber. Mei-P26 expression increases during cystoblast transit, as indicated. Expression levels of selected proliferation markers (Cyclin E, d-Myc) that correlate with Mei-P26 are shown. In *mei-P26* loss of function mutants cystocyte size is undiminished as the cells hyperproliferate. Differentiation to oocyte and nurse cells is disrupted, with formation of cystic egg chambers. Cyclin E and d-Myc are deregulated. d) A model for Trim32 in the neuroepithelium is depicted. Two radial progenitors (NPC a and b) undergoing interkinetic nuclear migration are shown, with apical feet and basal processes. At division Trim32 accumulates near the thinned basal process, to be differentially inherited by an incipient neuron (N). Expression of Trim32, let-7 and c-Myc in the NPC and neural daughter cells are indicated. A color version of this figure is available at www.landesbioscience.com/curie.

execute specific programs of proliferation and specification, but the basic scenario of asymmetric division is that the apical daughter retains neuroblast character while the basal daughter, called a ganglion mother cell (GMC), is smaller and destined for cell cycle exit and terminal differentiation. Segregation of cell fate determinants requires the action of apical proteins in the mother cell, including Par-3, Par-6 and atypical Protein Kinase C (reviewed in refs. 69,70)(Fig 3). These proteins are required for basal accumulation of a second set of proteins, which drive GMC specification. Brat is a recent entry into this group of specification factors, which includes Numb, an inhibitor of the Notch pathway, and Prospero, a transcription factor that suppresses self-renewal and promotes neural differentiation.[67,71,72] Brat was shown to bind Miranda,[71,72] a scaffold protein that accumulates in a crescent at the basal pole of the mother cell destined for inheritance by the GMC (or basal neuroblast, see below). After division, Brat suppresses neuroblast markers and promotes cell cycle exit in the GMC.[67,71,72] One proposed regulatory target for Brat is the *Drosophila* Myc protein, a neuroblast marker that is dysregulated in *brat* mutants.[71] A new wrinkle in this basic model comes from the discovery that the tumor lineage in *brat* mutants does not immediately produce a GMC but instead a secondary neuroblast[68] (Fig. 3b). This is a transit-amplifying cell that produces GMCs after further division. In this scenario, loss of Brat leads to tumors because the secondary neuroblast fails to mature. How Brat controls proliferation and maturation of these cells remains unclear, but it appears to work cooperatively and in parallel to Numb.[68] Recent evidence derived from studies of Mei-P26[65] and other Trim-NHLs to be discussed next points to a role for Brat as regulator of the miRNA pathway, but this has yet to be demonstrated.

TRIM-NHL PROTEINS AS REGULATORS OF THE miRNA PATHWAY

Mei-P26

Several of the biological processes relevant to the *Drosophila mei-P26* gene have already been mentioned in relation to other Trim-NHL proteins: seizure (Trim2), gonad development (LIN-41) and centrosome function (Trim32). The *mei-P26* gene was first encountered as recombination defective in a screen for mutants in meiosis.[73] Further analysis extended the phenotype of *mei-P26* mutants to both male and female sterility.[74] Affected ovaries of the mutants contained inappropriately high numbers of poorly differentiated cells. Testes were cystic and failed to produce motile spermatozoa.[74] Despite strong specificity for gonads in its zygotic expression, RNAi against *mei-P26* resulted in disruption of embryonic CNS and PNS organization.[75] Supporting a role in the CNS, *mei-P26* was picked up in a mutational screen as a strong suppressor of seizures in an *eas* (*ea*sily *s*hocked) background.[76] *eas* encodes an ethanolamine kinase; the exact nature of seizure susceptibility is not known but the *eas* mutation affects phospholipid metabolism and neuroblast proliferation.[77,78] Two missense mutations were discovered in the *mei-P26* suppressor allele, one in the coiled-coil domain and one in the NHL domain.[76]

Little information on the molecular function of Mei-P26 was available until 2008, when the Knoblich laboratory first implicated Trim-NHL proteins in the regulation of the miRNA pathway.[65] In visually stunning work, they first carefully dissected the growth control failure in the ovaries of *mei-P26* mutants. Terminal differentiation to oocytes was blocked in *mei-P26* ovaries, with an accumulation of cells in an intermediate state of differentiation in which proliferation markers (CyclinE, Phospho-histone-H3) were upregulated (see Fig. 3c). In analogy to their earlier work on Brat, Myc expression and nucleolar size were increased in the germ cell progeny of *mei-P26* mutants. Conversely, overexpression of Mei-P26 led to depletion of ovarian germ cells. Mei-P26 expression was upregulated in the initial progeny of ovarian stem cells and Mei-P26 levels increased during the four rounds of transit amplifying proliferative cell divisions. High Mei-P26 levels direct cell cycle exit and entry into terminal differentiation (Fig. 3c).

The model for Mei-P26 function in the germ line developed by Neumüller et al. shares similarity to the proposed role of LIN-41 in the vulva and seam cells[4] and of Brat in neuroblasts.[67,68,71,72] Support for a common mechanism came from the observation that Mei-P26 and Brat interact with Ago1,[65] the *Drosophila* Argonaute family member responsible for miRNA-mediated gene silencing.[79] The physical link between Mei-P26 and Ago1 was reflected in global miRNA expression levels: increased in *mei-P26* mutants and decreased upon Mei-P26 overexpression.[65] On balance, miRNAs appear to support ovarian stem cell self-renewal, based on the phenotypes of mutations that block miRNA biogenesis in the ovary.[80,81] miRNA function was also affected, as Mei-P26 interfered with repression of target genes by the *bantam* miRNA. Suppression of *bantam* by Mei-P26 may be a significant part of the pathway, as *bantam* is required for germ cell self-renewal and *mei-P26* is itself a predicted target for *bantam*. These results were the first indication that regulatory interactions between miRNAs and Trim-NHL proteins are a two-way street, which will be the theme of the rest of the review.

NHL-2

In a recent paper, the group of Victor Ambros set out to test if LIN-41 and other Trim-NHL proteins share functions in the *C. elegans* developmental timing pathway.[82] Mutations in *nhl-2* led to premature stem cell maturation in a manner reminiscent of *lin-4* and *let-7* family mutants (*let-7, mir-48, mir-84, mir-241*). Combining mutations in *let-7* family members with the *nhl-2* mutant exacerbated the phenotype, as would be expected if NHL-2 and *let-7* cooperate. Furthermore, cooperativity was not limited to *let-7* but was also seen for the unrelated miRNA *lsy-6*. Interestingly, loss of *nhl-2* was found to partially rescue a *lin-41* allele, suggesting that NHL-2 and LIN-41 might be actors in a single pathway, but with opposite roles. The demonstration that NHL-2 interacts with the CGH-1 protein in a two-hybrid screen strengthened the link to the miRNA pathway.[82] CGH-1 is homologous with the human RCK/p54 protein, a DEAD Box helicase that associates with Argonautes and the miRISC in P-bodies.[83] NHL-2 was then shown to colocalize to P-bodies and to physically and genetically interact with the core miRISC proteins ALG-1 and AIN-1 (*C. elegans* Argonaute and GW182 orthologs, respectively). Despite the ability to enhance silencing of several miRNA target genes, NHL-2 did not increase miRNA expression levels, suggesting that NHL-2 enhances the efficiency of the miRISC downstream of miRNA biogenesis.[82]

Following the demonstration that Mei-P26 is an inhibitor of the miRNA pathway,[37] the work by Hammell et al. showed that NHL-2 has the opposite role as a positive regulator (reviewed in ref. 84). Mechanistically, the action of the two proteins has not yet been clarified, but several models for NHL-2 were discussed.[82] Relying on the precedent for Brat as scaffold protein and the finding that the interaction between NHL-2 and the miRISC is RNase sensitive, one possibility is that NHL-2 might facilitate binding of CGH-1 to the miRISC effector-mRNA complex. This is consistent with the role of yeast and human orthologs of CGH-1 in promoting mRNA decapping and translational repression.[85,86]

Alternatively, action of NHL-2 as E3 ubiquitin ligase might serve to modify protein-protein interactions or activities within the miRISC. One interesting suggestion is that NHL-2 might fulfill a fail-safe function by ubiquitinating and promoting the degradation of any translation products that escape translational silencing. Insights into the mechanism of NHL-2 function should soon emerge, perhaps through studies that combine the power of *C. elegans* and *D. melanogaster* genetics with in vitro assays for miRISC activity.

Trim32 and the miRNA Pathway

Published back-to-back with the NHL-2 paper, the Knoblich lab reported that mouse Trim32 also functions as an activator of the miRNA pathway.[37] Pursuing functional homologs of Brat in mammalian neurogenesis, Trim32 was first shown to suppress proliferation of heterologous cells. During cortical development, expression of Trim32 tracked cell cycle exit and neuronal differentiation both temporally and spatially. Like Brat, Trim32 was shown to distribute asymmetrically to the daughter cells after some, but not all, neural progenitor cell divisions. Asymmetric inheritance was most frequent at the high point of neurogenesis around E14.5 (see Fig. 3d). Consistent with this, cultured neural progenitors displayed symmetric Trim32 inheritance during proliferative divisions but switched to an asymmetric mode under conditions favoring neurogenesis. These findings suggest that neurogenic signals regulate Trim32 localization, although the nature of the signal or the mechanism causing asymmetric Trim32 accumulation has not been determined. Current discussions of the determinants of asymmetry and the relationship to cell fate in the mammalian neuroepithelium are complex and controversial and are beyond the scope of this review (refer to refs. 69,87,88). One clue was that Trim32 concentrates in the retracting basal fiber (also referred to as radial fiber or basal process), a structural feature of mammalian neural progenitors that is involved in the initiation of the cleavage furrow and perhaps asymmetric daughter cell fate (Fig. 3d) (reviewed in ref. 89).

To better characterize the role of Trim32 in neural progenitors, Schwamborn et al. first combined in utero electroporation to manipulate Trim32 expression levels in the neuroepithelium in vivo with subsequent in vitro culture of the transfected cells to follow their fate.[37] Trim32 overexpression reduced the proliferative capacity and enhanced neuronal marker expression of the transfected cells. Conversely, Trim32 knockdown led to enhanced proliferation and delay in in vitro neurogenesis. Similar results were obtained when cell fate was examined in vivo: after electroporation. Trim32 overexpressing cells showed an increased rate of migration into the cortical layers; Trim32 knockdown cells were delayed in their exit from the progenitor zone. These results support a model in which Trim32 plays a similar role to Brat: asymmetric Trim32 inheritance might support cell cycle exit and favor neuronal fate choice.

To extend the parallel to Brat, Schwamborn et al. next demonstrated that Trim32 inhibits c-Myc. c-Myc protein levels were reduced in cells overexpressing Trim32, accompanied by accumulation of polyubiquitinated c-Myc. Both effects were eliminated after mutation of the RING domain in Trim32, suggesting that Trim32 might directly target c-Myc for degradation. A second parallel to Brat (and Mei-P26) was then explored, with the demonstration that Trim32 interacts with Ago1 in immunoprecipitation assays. Comparison of the signature of miRNAs that coprecipitate with Trim32 or Ago1 identified eight miRNAs with significant enrichment in Trim32 complexes. One of the significant hits was let-7a (but not other let-7 family members). To test a role for let-7a in neuronal differentiation, let-7a activity was blocked by in utero delivery of an anti-let-7a LNA antagomir. Cells receiving the let-7a antagomir had an approximately two-fold higher likelihood of retaining the progenitor marker Nestin. Conversely, overexpression of let-7a in the progenitor zone led to a strong increase in neuronal differentiation, as indicated by neuronal marker expression. This result suggests that Trim32 acts in part by increasing the activity of let-7a in neural progenitors.[37]

Like many landmark papers, the work of Schwamborn and coworkers raises as many new questions as it answers. Given the central role that c-Myc and let-7 play in the circuitry of stem cells, with Myc acting as a suppressor of let-7 transcription while also being a direct target gene for let-7 (reviewed in ref. 90, see also ref. 91), it is unclear if the regulatory interactions with Myc and let-7 represent independent activities of Trim32. It would be interesting to confirm if c-Myc and Trim32 directly associate and in what cellular compartment the interaction takes place. Does Trim32 draw Myc into P-bodies, as has been suggested for Piasy?[36] It would also be interesting to know if the effect of Trim32 on c-Myc degradation extends to other Myc family members, including N-Myc, a protein more directly tied to neurogenesis than c-Myc.[92,93] Also, if downregulation of Myc is central to both Brat and Trim32, is the RING-less Brat able to indirectly mediate c-Myc degradation? This question can be stated more broadly, since it is unclear if the E3 activity of Trim32, as well as NHL-2, is required for enhancing miRNA pathway activity. Schwamborn et al. reported that the RING domain of Trim32 was not required to suppress let-7 activity in a reporter assay, but it is far from clear if c-Myc is the only relevant target for Trim32 in suppressing cell proliferation and encouraging neuronal differentiation in vivo.[37] Another intriguing possibility is whether or not asymmetric Brat and Trim32 segregation to daughter cells implies differential inheritance of miRNAs and the core miRISC machinery. Does Trim32 pull neurogenic miRNAs and an activated miRISC into the cell destined to adopt a neuronal fate? It would be very interesting to compare the activity of reporters for Trim32-associated miRNAs before and after symmetrical and asymmetrical cell divisions. Such a mechanism might have functional relevance, as several studies reported that neural progenitors express let-7 prior to terminal differentiation.[94-96] Finally, it is not yet known if the Trim32-Ago1 interaction is unique or common to all Argonautes. This would be important in assessing the significance of preferential sorting of miRNAs to miRNPs containing Trim32. It is not currently known how functionally redundant mammalian Argonautes are and if they favor some miRNAs over others. Therefore, loading of miRNAs to the Trim32-Ago complex might be a consequence of which Ago is bound to Trim32, or to the ability of Trim32 to alter the properties of the miRISC.

mLin41

Based on mouse and human genetics, as discussed above, Trim2 and Trim32 are developmental regulators of the mammalian CNS. Mouse LIN-41 (mLin41) is the third mammalian Trim-NHL protein shown by genetic means to be involved in CNS morphogenesis.[97] Homozygous disruption of mLin41 in gene-trap lines revealed that mLin41 is required for embryonic viability, with a striking but uncharacterized failure of neural tube closure. Using the gene-trap as a reporter for mLin41 promoter activity in heterozygotes, strong expression was detected in the brain, dorsal root ganglia, eyes and branchial arches (and elsewhere) between E10.5 and E12.5. This corresponds closely to the period of embryonic demise in the homozygotes.[97] However, it is unclear if the craniofacial abnormalities and closure defect is specific or if it is a secondary result of growth arrest or reduced embryonic viability. Probing mLin41 expression at the protein level, we confirmed mLin41 expression in embryonic stem cells and in Oct-4[+] cells of the E7 embryonic ectoderm.[23] After embryonic induction of let-7 and miR-125 combine to downregulate *mLin41*, postnatal niches of mLin41 expression were found in ciliated epithelia of the male and female reproductive tract, in male germ cells and in interfollicular epidermal stem cells. Löer et al. had previously demonstrated the mLin41 protein in muscle[53] and, thus, the expression pattern of mLin41 seems quite analogous to that of the *C. elegans* protein.[4]

In pluripotent cells and in transfection assays in heterologous cells, mLin41 was found to accumulate in cytoplasmic P-bodies, as confirmed by colocalization with the P-body markers Dcp1a and Hedls as well as the miRISC proteins Ago2, Mov10 and Tnrc6b.[23] This observation dovetailed with a report that the *C. elegans* protein copurifies with the Dicer protein.[98] Association with Dicer was confirmed for mLin41 and extended to Ago2.[23] The interaction surface was mapped to the coiled-coil domain, suggesting the NHL domain is free to participate in additional protein-protein interactions. Although the association with Ago2 was studied in greater detail, cotransfection assays with tagged proteins confirmed interaction with Ago1 and Ago4. mLin41 is thus the sixth Trim-NHL protein shown to interact with one of the Argonautes (Mei-P26, Brat, Dappled,[65] NHL-2[82] and Trim32[37]). Unlike the others, however, mLin41 was shown to mediate degradative ubiquitination of Ago2.[23] First, in an autoubiquitination assay, E3 ubiquitin ligase activity was confirmed, with the same E2 preference displayed by Trim2 (UbcH5a).[22] Adding immunoprecipitated Ago2 to the assay led to the accumulation of polyubiquitinated Ago2. In a transfection assay, Ago2 ubiquitination was dependent on an intact RING domain and was strongly enhanced in the presence of the proteasome inhibitor MG132. In keeping with these results, depletion of endogenous mLin41 in embryocarcinoma cells decreased Ago2 ubiquitination and increased steady state Ago2 levels.[23]

If mLin41 regulates Ago2 turnover and potentially other miRISC components, it should also act as a general inhibitor of miRNA-mediated silencing. Two observations support such a role. Argonautes are thought to be limiting for miRNA-mediated silencing and ectopic Ago2 expression enhances miRNA accumulation,[99] most likely by protecting small RNAs from ribonucleolytic degradation.[100] mLin41 was shown to block the ability of Ago2 to increase let-7 levels.[23] It will be interesting to see if a similar mechanism is responsible for the global reduction in miRNA levels observed after overexpression of Mei-P26.[65] Furthermore, mLin41 reduced silencing mediated by let-7, miR-124 or miR-128 in reporter assays. Interference with miRNA activity was dependent on an intact RING domain and could be compensated by increased expression of Ago2.[23] This finding was expanded by demonstrating that mLin41 cooperated with Lin28 in suppressing let-7 in the reporter assay, providing evidence for dual negative autoregulatory loops in the post-transcriptional control of let-7 in pluripotent cells (see Fig. 4a and 4b for a schematic view).

CONCLUSION

Trim-NHL proteins serve as pleiotropic regulators of developmental processes and cell function. In development, a picture is emerging in which Trim-NHL proteins coordinate miRNA activity to sequentially drive transitions between self-renewal, commitment and terminal differentiation of stem cells. The picture is incomplete, in part because recent advances have come from parallel studies in *C. elegans*, *D. melanogaster* and mammals so

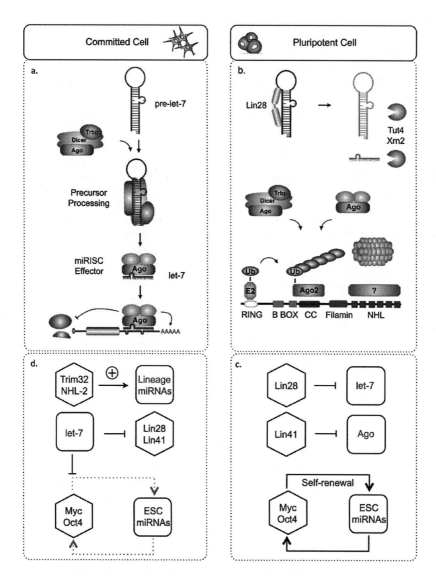

Figure 4. Summary of miRNA regulation during stem cell differentiation. a) In committed cells such as neural progenitors and their offspring, miRNA precursors such as pre-let-7 are processed by a core complex of Dicer, Trbp and one of the Argonaute proteins (Ago). After cleavage, one strand is passed to an effector complex. Composition of the effector is not yet fully defined, but is centered on one Ago and associated proteins such as GW182/Tnrc6, Mov10, FMRP and Rck/p54 (in mammals). The effector mediates translational silencing and frequently enhanced mRNA decay. b) In pluripotent cells, the pre-let-7 specific RNA-binding protein Lin28 blocks access to pre-let-7, which is thereby subject to uridylation and enhanced degradation. Lin41 binds Ago via the coiled-coil domain, stimulating ubiquitin-mediated proteolysis. c) In pluripotent cells, let-7 maturation is inhibited by Lin28, while Ago (and perhaps additional miRISC proteins) are downregulated by Lin41. A limited class of ES-specific miRNAs participates in a positive feedback loop with pluripotency-associated transcription factors (Myc, Nanog, Oct-4, Sox2). d) Activation of let-7 results in suppression of Lin28 and Lin41. Let-7 inhibits the transcriptional program of pluripotent cells and downregulates ESC miRNAs. Trim32 (or NHL-2 in *C. elegans*) act to enhance miRNA activity. The signal that releases let-7 repression is not known, but presumably involves Myc and miR-125. See text and references for details.

that the issues of sequential action and redundancy have not yet been thoroughly addressed for any one developmental context. For example, in *C. elegans,* LIN-41 and NHL-2 have opposite roles in the heterochronic pathway governing stem cell maturation in the hypodermis and vulva.[82] Assuming that LIN-41 is an inhibitor of miRNA pathway activity, as has been shown for the mouse protein, then LIN-41 most likely blocks the acquisition of adult cell fates in the hypodermis at least in part by inhibiting full let-7 activity (the primary miRNA driver of maturation in progenitor populations). Since LIN-41 is itself a *let-7* target,[4] this is yet another example of a double negative loop enhancing the potency of a miRNA-target gene interaction (see Fig. 4). By promoting miRNA activity, NHL-2 can flip this switch and unleash the full effects of *let-7* on downstream heterochronic regulators.

Specification of mouse embryonic stem cells (ESCs) appears to proceed by a similar mechanism. In undifferentiated, pluripotent cells, self-renewal is reinforced by a regulatory loop comprising a unique class of ESC-specific miRNAs and transcription factors such as Oct-4, Nanog and Myc.[91,101] In ESCs, let-7 is inhibited by the combined action of Lin28 and Lin41[23] (reviewed in ref. 102 and in the chapter by Lehrbach and Miska). The importance of let-7 is threefold. First, the impact of let-7 on the proteome is amplified due to direct targeting of miRNA pathway genes.[23,96,103-105] Second, let-7 targets cell cycle regulators, thereby directly influencing proliferation.[106] Third, let-7 suppresses the transcriptional program of transcription factors required for the maintenance of pluripotency and self-renewal by both direct (c-Myc, N-Myc) and indirect mechanisms (Oct-4, Nanog, Sox2, Tcf3).[91] By analogy to the opposing action of LIN-41 and NHL-2 in *C. elegans*, it is reasonable to expect a counterpoint to mLin41 which activates the miRNA pathway during commitment. In neural lineages, Trim32 fulfills two of the expected criteria: it activates a select class of miRNAs, including let-7a, and suppresses c-Myc.[37]

The dual function exerted by at least some Trim-NHLs in the ubiquitin and miRNA pathways should provide a framework for exploring their functional pleiotropy. It will be important to determine whether miRISC association is common to all family members. Within the miRISC, evidence from a variety of sources suggests that Ago2 is not the only substrate for ubiquitination.[23,107,108] The ubiquitination of additional miRISC components could certainly affect more than just the turnover rate of the individual proteins in the complex. Complex assembly, activity and intracellular compartmentalization might also be influenced. One obvious possibility is that the ability of many Trim-NHLs to associate with cytoskeletal components in a variety of cellular compartments may reflect a role in miRISC transport.[108-111] Alternatively, the association of Wech with focal adhesions or Trim3 with CART may be evidence of miRISC-independent activities.

A recent study directly assayed functional redundancy and cooperation among the five Trim-NHLs in *C. elegans.* Mutations in four of the five (with the exception of LIN-41) were able to suppress a mutation in the embryonic polarity gene *par-2,* suggesting substantial functional redundancy in this pathway.[10] However, other specific functions differed[10] and the function of NHL-2 in the heterochronic pathway later in development appears to be independent of NCL-1, NHL-1 or NHL-3.[82] In the mouse, loss of Lin41 cannot be compensated and leads to embryonic lethality.[97] On the other hand, no obvious defect in neurogenesis was reported after deletion of *Trim32,*[44] in contrast to the direct assays of neuronal differentiation reported by Schwamborn et al.[37] In this case Trim2 and Trim3 may functionally compensate for Trim32 deficiency. Coiled-coil domains can mediate heteromeric interactions,[9] but there is no evidence yet for cooperativity among Trim-NHLs or between Trim-NHLs and other E3 ligases. If Trim-NHLs can form heterodimers, then the RING-less proteins Brat and Dappled/Wech might promote ubiquitination indirectly.

The discovery of ubiquitin-mediated regulation of miRNAs has revealed a nexus between two of the cell's most powerful post-transcriptional regulatory pathways. The miRNA pathway appears to make extensive use of autoregulatory feedback loops to adjust miRNA activity during development and stem cell differentiation. Several of the key molecules in these loops are involved in tumor formation (Brat in *Drosophila* or let-7, Lin28 and Myc in humans), underscoring their potential relevance for human disease. Two developmental disturbances have been linked to Trim32, more may be uncovered as the other human Trim-NHLs are studied. In addition to development, regulation of Trim-NHL protein synthesis or activity, for example in response to cell signaling, may allow cells to modulate miRNA efficiency to maintain cellular homeostasis.

ACKNOWLEDGEMENTS

F.G.W., E.F. and A.R. received support from the DFG Collaborative research project 665; E.C. and A.R were supported by the DFG Graduate School 1123.

REFERENCES

1. Reinhart BJ, Slack FJ, Basson M et al. The 21-nucleotide let-7 RNA regulates developmental timing in Caenorhabditis elegans. Nature 2000; 403(6772):901-906.
2. Pasquinelli AE, Reinhart BJ, Slack F et al. Conservation of the sequence and temporal expression of let-7 heterochronic regulatory RNA. Nature 2000; 408(6808):86-89.
3. Lee R, Feinbaum R, Ambros V. A short history of a short RNA. Cell 2004; 116(2 Suppl):S89-92, 81.
4. Slack FJ, Basson M, Liu Z et al. The lin-41 RBCC gene acts in the C. elegans heterochronic pathway between the let-7 regulatory RNA and the LIN-29 transcription factor. Mol Cell 2000; 5(4):659-669.
5. Slack FJ, Ruvkun G. A novel repeat domain that is often associated with RING finger and B-box motifs. Trends Biochem Sci 1998; 23(12):474-475.
6. Meroni G, Diez-Roux G. TRIM/RBCC, a novel class of 'single protein RING finger' E3 ubiquitin ligases. Bioessays 2005; 27(11):1147-1157.
7. Sardiello M, Cairo S, Fontanella B et al. Genomic analysis of the TRIM family reveals two groups of genes with distinct evolutionary properties. BMC Evol Biol 2008; 8:225.
8. Nisole S, Stoye JP, Saib A. TRIM family proteins: retroviral restriction and antiviral defence. Nat Rev Microbiol 2005; 3(10):799-808.
9. Reymond A, Meroni G, Fantozzi A et al. The tripartite motif family identifies cell compartments. EMBO J 2001; 20(9):2140-2151.
10. Hyenne V, Desrosiers M, Labbe JC. C. elegans Brat homologs regulate PAR protein-dependent polarity and asymmetric cell division. Dev Biol 2008; 321(2):368-378.
11. O'Farrell F, Esfahani SS, Engstrom Y et al. Regulation of the Drosophila lin-41 homologue dappled by let-7 reveals conservation of a regulatory mechanism within the LIN-41 subclade. Dev Dyn 2008; 237(1):196-208.
12. Saurin AJ, Borden KL, Boddy MN et al. Does this have a familiar RING? Trends Biochem Sci 1996; 21(6):208-214.
13. Deshaies RJ, Joazeiro CA. RING domain E3 ubiquitin ligases. Annu Rev Biochem 2009; 78:399-434.
14. Joazeiro CA, Wing SS, Huang H et al. The tyrosine kinase negative regulator c-Cbl as a RING-type, E2-dependent ubiquitin-protein ligase. Science 1999; 286(5438):309-312.
15. Yokouchi M, Kondo T, Houghton A et al. Ligand-induced ubiquitination of the epidermal growth factor receptor involves the interaction of the c-Cbl RING finger and UbcH7. J Biol Chem 1999; 274(44):31707-31712.
16. Lorick KL, Jensen JP, Fang S et al. RING fingers mediate ubiquitin-conjugating enzyme (E2)-dependent ubiquitination. Proc Natl Acad Sci USA 1999; 96(20):11364-11369.
17. Freemont PS. RING for destruction? Curr Biol 2000; 10(2):R84-87.
18. Lupas A. Coiled coils: new structures and new functions. Trends Biochem Sci 1996; 21(10):375-382.
19. Zheng N, Wang P, Jeffrey PD et al. Structure of a c-Cbl-UbcH7 complex: RING domain function in ubiquitin-protein ligases. Cell 2000; 102(4):533-539.
20. Horn EJ, Albor A, Liu Y et al. RING protein Trim32 associated with skin carcinogenesis has anti-apoptotic and E3-ubiquitin ligase properties. Carcinogenesis 2004; 25(2):157-167.

21. Kudryashova E, Kudryashov D, Kramerova I et al. Trim32 is a ubiquitin ligase mutated in limb girdle muscular dystrophy type 2H that binds to skeletal muscle myosin and ubiquitinates actin. J Mol Biol 2005; 354(2):413-424.

22. Balastik M, Ferraguti F, Pires-da Silva A et al. Deficiency in ubiquitin ligase TRIM2 causes accumulation of neurofilament light chain and neurodegeneration. Proc Natl Acad Sci USA 2008; 105(33):12016-12021.

23. Rybak A, Fuchs H, Hadian K et al. The let-7 target gene mouse lin-41 is a stem cell specific E3 ubiquitin ligase for the miRNA pathway protein Ago2. Nat Cell Biol 2009; 11(12):1411-1420.

24. Chiang AP, Beck JS, Yen HJ et al. Homozygosity mapping with SNP arrays identifies TRIM32, an E3 ubiquitin ligase, as a Bardet-Biedl syndrome gene (BBS11). Proc Natl Acad Sci USA 2006; 103(16):6287-6292.

25. Saccone V, Palmieri M, Passamano L et al. Mutations that impair interaction properties of TRIM32 associated with limb-girdle muscular dystrophy 2H. Hum Mutat 2008; 29(2):240-247.

26. Massiah MA, Matts JA, Short KM et al. Solution structure of the MID1 B-box2 CHC(D/C)C(2)H(2) zinc-binding domain: insights into an evolutionarily conserved RING fold. J Mol Biol 2007; 369(1):1-10.

27. Massiah MA, Simmons BN, Short KM et al. Solution structure of the RBCC/TRIM B-box1 domain of human MID1: B-box with a RING. J Mol Biol 2006; 358(2):532-545.

28. Tao H, Simmons BN, Singireddy S et al. Structure of the MID1 tandem B-boxes reveals an interaction reminiscent of intermolecular ring heterodimers. Biochemistry 2008; 47(8):2450-2457.

29. Grigoryan G, Keating AE. Structural specificity in coiled-coil interactions. Curr Opin Struct Biol 2008; 18(4):477-483.

30. Zhou AX, Hartwig JH, Akyurek LM. Filamins in cell signaling, transcription and organ development. Trends Cell Biol.

31. Arama E, Dickman D, Kimchie Z et al. Mutations in the beta-propeller domain of the Drosophila brain tumor (brat) protein induce neoplasm in the larval brain. Oncogene 2000; 19(33):3706-3716.

32. Sonoda J, Wharton RP. Drosophila brain tumor is a translational repressor. Genes Dev 2001; 15(6):762-773.

33. Edwards TA, Wilkinson BD, Wharton RP et al. Model of the brain tumor-Pumilio translation repressor complex. Genes Dev 2003; 17(20):2508-2513.

34. Pickart CM, Cohen RE. Proteasomes and their kin: proteases in the machine age. Nat Rev Mol Cell Biol 2004; 5(3):177-187.

35. Schulman BA, Harper JW. Ubiquitin-like protein activation by E1 enzymes: the apex for downstream signalling pathways. Nat Rev Mol Cell Biol 2009; 10(5):319-331.

36. Albor A, El-Hizawi S, Horn EJ et al. The interaction of Piasy with Trim32, an E3-ubiquitin ligase mutated in limb-girdle muscular dystrophy type 2H, promotes Piasy degradation and regulates UVB-induced keratinocyte apoptosis through NFkappaB. J Biol Chem 2006; 281(35):25850-25866.

37. Schwamborn JC, Berezikov E, Knoblich JA. The TRIM-NHL protein TRIM32 activates microRNAs and prevents self-renewal in mouse neural progenitors. Cell 2009; 136(5):913-925.

38. Locke M, Tinsley CL, Benson MA et al. TRIM32 is an E3 ubiquitin ligase for dysbindin. Hum Mol Genet 2009; 18(13):2344-2358.

39. Fridell RA, Harding LS, Bogerd HP et al. Identification of a novel human zinc finger protein that specifically interacts with the activation domain of lentiviral Tat proteins. Virology 1995; 209(2):347-357.

40. Frosk P, Weiler T, Nylen E et al. Limb-girdle muscular dystrophy type 2H associated with mutation in TRIM32, a putative E3-ubiquitin-ligase gene. Am J Hum Genet 2002; 70(3):663-672.

41. Li W, Zhang Q, Oiso N et al. Hermansky-Pudlak syndrome type 7 (HPS-7) results from mutant dysbindin, a member of the biogenesis of lysosome-related organelles complex 1 (BLOC-1). Nat Genet 2003; 35(1):84-89.

42. Mykytyn K, Sheffield VC. Establishing a connection between cilia and Bardet-Biedl Syndrome. Trends Mol Med 2004; 10(3):106-109.

43. Ansley SJ, Badano JL, Blacque OE et al. Basal body dysfunction is a likely cause of pleiotropic Bardet-Biedl syndrome. Nature 2003; 425(6958):628-633.

44. Kudryashova E, Wu J, Havton LA et al. Deficiency of the E3 ubiquitin ligase TRIM32 in mice leads to a myopathy with a neurogenic component. Hum Mol Genet 2009; 18(7):1353-1367.

45. Pasquinelli AE, Ruvkun G. Control of developmental timing by micromas and their targets. Annu Rev Cell Dev Biol 2002; 18:495-513.

46. Kloosterman WP, Wienholds E, Ketting RF et al. Substrate requirements for let-7 function in the developing zebrafish embryo. Nucleic Acids Res 2004; 32(21):6284-6291.

47. Lancman JJ, Caruccio NC, Harfe BD et al. Analysis of the regulation of lin-41 during chick and mouse limb development. Dev Dyn 2005; 234(4):948-960.

48. Schulman BR, Esquela-Kerscher A, Slack FJ. Reciprocal expression of lin-41 and the microRNAs let-7 and mir-125 during mouse embryogenesis. Dev Dyn 2005; 234(4):1046-1054.

49. Kanamoto T, Terada K, Yoshikawa H et al. Cloning and regulation of the vertebrate homologue of lin-41 that functions as a heterochronic gene in Caenorhabditis elegans. Dev Dyn 2006; 235(4):1142-1149.

50. Rodriguez A, Zhou Z, Tang ML et al. Identification of immune system and response genes and novel mutations causing melanotic tumor formation in Drosophila melanogaster. Genetics 1996; 143(2):929-940.
51. Brody T, Stivers C, Nagle J et al. Identification of novel Drosophila neural precursor genes using a differential embryonic head cDNA screen. Mech Dev 2002; 113(1):41-59.
52. O'Farrell F, Kylsten P. A mis-expression study of factors affecting Drosophila PNS cell identity. Biochem Biophys Res Commun 2008; 370(4):657-662.
53. Loer B, Bauer R, Bornheim R et al. The NHL-domain protein Wech is crucial for the integrin-cytoskeleton link. Nat Cell Biol 2008; 10(4):422-428.
54. Delon I, Brown N. Cell-matrix adhesion: the wech connection. Curr Biol 2008; 18(9):R389-391.
55. Loer B, Hoch M. Wech proteins: roles in integrin functions and beyond. Cell Adh Migr 2008; 2(3):177-179.
56. Sokol NS, Xu P, Jan YN et al. Drosophila let-7 microRNA is required for remodeling of the neuromusculature during metamorphosis. Genes Dev 2008; 22(12):1591-1596.
57. El Husseini AE, Vincent SR. Cloning and characterization of a novel RING finger protein that interacts with class V myosins. J Biol Chem 1999; 274(28):19771-19777.
58. Trybus KM. Myosin V from head to tail. Cell Mol Life Sci 2008; 65(9):1378-1389.
59. Yan Q, Sun W, Kujala P et al. CART: an Hrs/actinin-4/BERP/myosin V protein complex required for efficient receptor recycling. Mol Biol Cell 2005; 16(5):2470-2482.
60. Mosesson Y, Chetrit D, Schley L et al. Monoubiquitinylation regulates endosomal localization of Lst2, a negative regulator of EGF receptor signaling. Dev Cell 2009; 16(5):687-698.
61. Zwang Y, Yarden Y. Systems biology of growth factor-induced receptor endocytosis. Traffic 2009; 10(4):349-363.
62. Ohkawa N, Kokura K, Matsu-Ura T et al. Molecular cloning and characterization of neural activity-related RING finger protein (NARF): a new member of the RBCC family is a candidate for the partner of myosin V. J Neurochem 2001; 78(1):75-87.
63. Gray PA, Fu H, Luo P et al. Mouse brain organization revealed through direct genome-scale TF expression analysis. Science 2004; 306(5705):2255-2257.
64. Wright TR. The Wilhelmine E. Key 1992 Invitational lecture. Phenotypic analysis of the Dopa decarboxylase gene cluster mutants in Drosophila melanogaster. J Hered 1996; 87(3):175-190.
65. Neumuller RA, Betschinger J, Fischer A et al. Mei-P26 regulates microRNAs and cell growth in the Drosophila ovarian stem cell lineage. Nature 2008; 454(7201):241-245.
66. Frank DJ, Edgar BA, Roth MB. The Drosophila melanogaster gene brain tumor negatively regulates cell growth and ribosomal RNA synthesis. Development 2002; 129(2):399-407.
67. Bello B, Reichert H, Hirth F. The brain tumor gene negatively regulates neural progenitor cell proliferation in the larval central brain of Drosophila. Development 2006; 133(14):2639-2648.
68. Bowman SK, Rolland V, Betschinger J et al. The tumor suppressors Brat and Numb regulate transit-amplifying neuroblast lineages in Drosophila. Dev Cell 2008; 14(4):535-546.
69. Knoblich JA. Mechanisms of asymmetric stem cell division. Cell 2008; 132(4):583-597.
70. Kohlmaier A, Edgar BA. Proliferative control in Drosophila stem cells. Curr Opin Cell Biol 2008; 20(6):699-706.
71. Betschinger J, Mechtler K, Knoblich JA. Asymmetric segregation of the tumor suppressor brat regulates self-renewal in Drosophila neural stem cells. Cell 2006; 124(6):1241-1253.
72. Lee CY, Wilkinson BD, Siegrist SE et al. Brat is a Miranda cargo protein that promotes neuronal differentiation and inhibits neuroblast self-renewal. Dev Cell 2006; 10(4):441-449.
73. Sekelsky JJ, McKim KS, Messina L et al. Identification of novel Drosophila meiotic genes recovered in a P-element screen. Genetics 1999; 152(2):529-542.
74. Page SL, McKim KS, Deneen B et al. Genetic studies of mei-P26 reveal a link between the processes that control germ cell proliferation in both sexes and those that control meiotic exchange in Drosophila. Genetics 2000; 155(4):1757-1772.
75. Ivanov AI, Rovescalli AC, Pozzi P et al. Genes required for Drosophila nervous system development identified by RNA interference. Proc Natl Acad Sci USA 2004; 101(46):16216-16221.
76. Glasscock E, Singhania A, Tanouye MA. The mei-P26 gene encodes a RING finger B-box coiled-coil-NHL protein that regulates seizure susceptibility in Drosophilia. Genetics 2005; 170(4):1677-1689.
77. Pavlidis P, Ramaswami M, Tanouye MA. The Drosophila easily shocked gene: a mutation in a phospholipid synthetic pathway causes seizure, neuronal failure and paralysis. Cell 1994; 79(1):23-33.
78. Pascual A, Chaminade M, Preat T. Ethanolamine kinase controls neuroblast divisions in Drosophila mushroom bodies. Dev Biol 2005; 280(1):177-186.
79. Okamura K, Ishizuka A, Siomi H et al. Distinct roles for argonaute proteins in small RNA-directed RNA cleavage pathways. Genes Dev 2004; 18(14):1655-1666.
80. Park JK, Liu X, Strauss TJ et al. The miRNA pathway intrinsically controls self-renewal of Drosophila germline stem cells. Curr Biol 2007; 17(6):533-538.
81. Jin Z, Xie T. Dcr-1 maintains Drosophila ovarian stem cells. Curr Biol 2007; 17(6):539-544.

82. Hammell CM, Lubin I, Boag PR et al. nhl-2 Modulates microRNA activity in Caenorhabditis elegans. Cell 2009; 136(5):926-938.
83. Chu CY, Rana TM. Translation repression in human cells by microRNA-induced gene silencing requires RCK/p54. PLoS Biol 2006; 4(7):e210.
84. Loedige I, Filipowicz W. TRIM-NHL proteins take on miRNA regulation. Cell 2009; 136(5):818-820.
85. Coller JM, Tucker M, Sheth U et al. The DEAD box helicase, Dhh1p, functions in mRNA decapping and interacts with both the decapping and deadenylase complexes. RNA 2001; 7(12):1717-1727.
86. Eulalio A, Rehwinkel J, Stricker M et al. Target-specific requirements for enhancers of decapping in miRNA-mediated gene silencing. Genes Dev 2007; 21(20):2558-2570.
87. Farkas LM, Huttner WB. The cell biology of neural stem and progenitor cells and its significance for their proliferation versus differentiation during mammalian brain development. Curr Opin Cell Biol 2008; 20(6):707-715.
88. Zhong W, Chia W. Neurogenesis and asymmetric cell division. Curr Opin Neurobiol 2008; 18(1):4-11.
89. Kosodo Y, Huttner WB. Basal process and cell divisions of neural progenitors in the developing brain. Dev Growth Differ 2009; 51(3):251-261.
90. Bussing I, Slack FJ, Grosshans H. let-7 microRNAs in development, stem cells and cancer. Trends Mol Med 2008; 14(9):400-409.
91. Melton C, Judson RL, Blelloch R. Opposing microRNA families regulate self-renewal in mouse embryonic stem cells. Nature 2010.
92. Knoepfler PS, Cheng PF, Eisenman RN. N-myc is essential during neurogenesis for the rapid expansion of progenitor cell populations and the inhibition of neuronal differentiation. Genes Dev 2002; 16(20):2699-2712.
93. Martins RA, Zindy F, Donovan S et al. N-myc coordinates retinal growth with eye size during mouse development. Genes Dev 2008; 22(2):179-193.
94. Kloosterman WP, Wienholds E, de Bruijn E et al. In situ detection of miRNAs in animal embryos using LNA-modified oligonucleotide probes. Nat Methods 2006; 3(1):27-29.
95. Nishino J, Kim I, Chada K et al. Hmga2 promotes neural stem cell self-renewal in young but not old mice by reducing p16Ink4a and p19Arf Expression. Cell 2008; 135(2):227-239.
96. Rybak A, Fuchs H, Smirnova L et al. A feedback loop comprising lin-28 and let-7 controls pre-let-7 maturation during neural stem-cell commitment. Nat Cell Biol 2008; 10(8):987-993.
97. Maller Schulman BR, Liang X, Stahlhut C et al. The let-7 microRNA target gene, Mlin41/Trim71 is required for mouse embryonic survival and neural tube closure. Cell Cycle 2008; 7(24):3935-3942.
98. Duchaine TF, Wohlschlegel JA, Kennedy S et al. Functional proteomics reveals the biochemical niche of C. elegans DCR-1 in multiple small-RNA-mediated pathways. Cell 2006; 124(2):343-354.
99. Diederichs S, Haber DA. Dual role for argonautes in microRNA processing and post-transcriptional regulation of microRNA expression. Cell 2007; 131(6):1097-1108.
100. Chatterjee S, Grosshans H. Active turnover modulates mature microRNA activity in Caenorhabditis elegans. Nature 2009; 461(7263):546-549.
101. Sinkkonen L, Hugenschmidt T, Berninger P et al. microRNAs control de novo DNA methylation through regulation of transcriptional repressors in mouse embryonic stem cells. Nat Struct Mol Biol 2008; 15(3):259-267.
102. Dueck A, Meister G. TRIMming microRNA function in mouse stem cells. Nat Cell Biol 2009; 11(12):1392-1393.
103. Selbach M, Schwanhausser B, Thierfelder N et al. Widespread changes in protein synthesis induced by microRNAs. Nature 2008; 455(7209):58-63.
104. Ding XC, Slack FJ, Grosshans H. The let-7 microRNA interfaces extensively with the translation machinery to regulate cell differentiation. Cell Cycle 2008; 7(19):3083-3090.
105. Tokumaru S, Suzuki M, Yamada H et al. let-7 regulates Dicer expression and constitutes a negative feedback loop. Carcinogenesis 2008.
106. Johnson CD, Esquela-Kerscher A, Stefani G et al. The let-7 microRNA represses cell proliferation pathways in human cells. Cancer Res 2007; 67(16):7713-7722.
107. Ashraf SI, McLoon AL, Sclarsic SM et al. Synaptic protein synthesis associated with memory is regulated by the RISC pathway in Drosophila. Cell 2006; 124(1):191-205.
108. Gibbings DJ, Ciaudo C, Erhardt M et al. Multivesicular bodies associate with components of miRNA effector complexes and modulate miRNA activity. Nat Cell Biol 2009; 11(9):1143-1149.
109. Kotaja N, Bhattacharyya SN, Jaskiewicz L et al. The chromatoid body of male germ cells: Similarity with processing bodies and presence of Dicer and microRNA pathway components. Proc Natl Acad Sci USA 2006.
110. Wulczyn FG, Smirnova L, Rybak A et al. Post-transcriptional regulation of the let-7 microRNA during neural cell specification. FASEB J 2007; 21(2):415-426.
111. Lee YS, Pressman S, Andress AP et al. Silencing by small RNAs is linked to endosomal trafficking. Nat Cell Biol 2009; 11(9):1150-1156.
112. Kirilly D, Xie T. The Drosophila ovary: an active stem cell community. Cell Res 2007; 17(1):15-25.

CHAPTER 10

PROPERTIES OF THE REGULATORY RNA-BINDING PROTEIN HuR AND ITS ROLE IN CONTROLLING miRNA REPRESSION

Nicole-Claudia Meisner* and Witold Filipowicz*

Abstract: Gene expression in eukaryotes is subject to extensive regulation at posttranscriptional levels. One of the most important sites of control involves mRNA 3′ untranslated regions (3'UTRs), which are recognized by RNA-binding proteins (RBPs) and microRNAs (miRNAs). These factors greatly influence translational efficiency and stability of target mRNAs and often also determine their cellular localization. HuR, a ubiquitously expressed member of the ELAV family of RBPs, has been implicated in regulation of stability and translation of over one hundred mRNAs in mammalian cells. Recent data indicate that some of the effects of HuR can be explained by its interplay with miRNAs. Binding of HuR may suppress the inhibitory effect of miRNAs interacting with the 3'UTR and redirect the repressed mRNA to polysomes for active translation. However, HuR can also synergize with miRNAs. The finding that HuR is able to disengage miRNAs from the repressed mRNA, or render them inactive, provides evidence that miRNA regulation is much more dynamic then originally anticipated. In this chapter we review properties of HuR and describe examples of the cross-talk between the protein and miRNAs, with emphasis on response of the regulation to cellular stress.

INTRODUCTION TO HuR AND ARE ELEMENTS

Post-transcriptional control had been recognized as a central mechanism in mammalian gene expression already in the preRNAi era, one of the most prominent pathways being governed by so called AU-rich elements (AREs).[1-3] These *cis* acting elements are usually found in 3'-untranslated regions from where they orchestrate control of mRNA turnover,

*Corresponding Authors: Witold Filipowicz—Friedrich Miescher Institute for Biomedical, Research, PO Box 2543, 4002 Basel, Switzerland. Email: witold.filipowicz@fmi.ch; and Nicole-Claudia Meisner—Novartis Institutes for Biomedical Research, Novartis Campus, 4002 Basel, Switzerland. Email: nicole-claudia.meisner@novartis.com

Regulation of microRNAs, edited by Helge Großhans.
©2010 Landes Bioscience and Springer Science+Business Media.

translation and/or transport by interacting with RNA-binding proteins (RBPs).[2,4,5] As for most other post-transcriptional processes, a hallmark of ARE-mediated regulation is a fast on- and off-response at the protein level. It is therefore no surprise to find that immediate-early and early response genes (i.e., genes which, in response to particular triggers such as growth factors, cytokines or changed environmental conditions, are activated transiently and rapidly, even before any new proteins are synthesized) in particular use this mechanism to rapidly change their expression profiles. Based on their sequence and decay-inducing characteristics, AREs were classified into three major types. As originally proposed by Chen and Shyu,[1,6] Class I AREs comprise multiple dispersed AUUUA pentamers. mRNAs bearing these AREs are characterized by a biphasic degradation kinetics, initiated by synchronous polyA-tail shortening prior to exonucleolytic processing of the mRNA body. mRNAs containing Class II AREs, which are defined by at least two overlapping copies of the AUUUA pentamer, have halflives usually shorter than that of Class I and III ARE mRNAs, and their turnover is characterized by asynchronous polyA-tail shortening, which is paralleled by decay of the mRNA body. Class III AREs comprise loosely defined non-AUUUA but U-rich elements and mRNAs containing them follow similar decay kinetics as Class I ARE mRNAs. With a current estimated existence of ~4000 ARE-controlled genes,[7] AREs are involved in diverse cellular processes, ranging from cell proliferation, migration, stress response, metabolism and cell signaling to differentiation and senescence (reviewed e.g., in refs. 1-3,8,9). Among the more than 20 ARE-binding proteins known to date, the majority has been associated with a role in promoting mRNA degradation (e.g., AUF1, BRF1, KSRP or TTP) and/or suppression of translation (e.g., TIAR, TIA-1).[2,10-12] The Hu family of RBPs are the most prominent antagonists of ARE-mediated downregulation. The Hu family includes four members: HuA (HuR, ANNA-1), HuB (Hel-N1), HuC (Ple-21) and HuD. HuR is ubiquitously expressed, while expression of the other three proteins is restricted to the neuronal system;[13,14] HuB was additionally reported to be expressed in gonads.[15]

Mammalian Hu proteins, similar to their Drosophila homologue, the ELAV (*embryonic lethal, abnormal vision*) protein, are composed of three RNA recognition motifs (RRM) (Fig. 1A). The two N-terminal RRM domains are most conserved across Hu family members and between different species and function in tandem to bind to the ARE.[8,16,17] While a general binding preference for U-rich sequences was noted early on,[16,17] the first evidence for a consensus motif came from crystallographic studies on HuD which showed binding to an 8mer of sequence NUUNNUUU.[18] An in vitro study using systematically designed synthetic short RNA ligands then derived a related motif for HuR showing a requirement for 9 nucleotides, NNUUNNUUU, in single-stranded conformation.[19] This means that such HuR binding sites are inherently present within Class I and Class II AREs, because NNUUNNUUU motifs are contained within their consensus sequences. However, the converse does not apply and an HuR binding motif is not sufficient to define a functional ARE. Therefore, we distinguish here between these two terms. An independent study used a more unbiased bioinformatics approach to identify a consensus of mRNAs immunoprecipitated with HuR based on primary and secondary structure analysis.[20] This approach delivered a less stringent motif of a 17-20 nucleotide stem-loop rich in uracils, which the authors validated by de novo prediction of further HuR targets. While these data suggest that this motif is a common feature of HuR targets, at this point it cannot be concluded whether it represents indeed a direct binding motif for HuR, particularly as it is in disagreement with the single stranded, AUUUA pentamer-related binding motif revealed by the biochemical studies.[18,19] In a paper describing IL-4 regulation by HuR,

Yarovinsky et al. reported that a computational analysis revealed no such stem-loop HuR consensus motif within murine IL-4 mRNA,[21] but rather two NNUUNNUUU motifs, both of which were confirmed to be at the core of sites complexed by HuR. A more recent study on the Eph receptor and ephrin 3'UTRs also identified HuR binding sites related to single stranded NNUUNNUUU motifs.[22] Next-generation methods for global analysis of RNA-protein interactions such as CLiP[23] or HITS-CLiP[24] are likely to further reveal identities of HuR binding sites within its target mRNAs.

In contrast to the highly conserved RRM1 and 2, the basic hinge region and C-terminal RRM3 show significant diversity among Hu proteins. The function of these domains remains less well characterized, although they are both associated with nucleocytoplasmic trafficking. HuR is a predominantly nuclear protein, but ARE-mediated decay is generally considered to occur in the cytoplasm. Hence, it was not surprising to find that HuR actively shuttles between the two compartments, as demonstrated by Fan and Steitz in 1998 through interspecies heterokaryon fusion experiments.[25] Shortly thereafter, the same group identified a unique nuclear localization signal within the hinge region of HuR which they termed HNS (HuR nuclear localization signal,[26] Fig. 1A). The basic 32-amino acid HNS mediates both nuclear localization and nuclear export of HuR. Gallouzi et al. later showed that HuR can use two shuttling pathways, one of which depends on CRM1.[27] The CRM1-dependent pathway does not involve HNS but depends on an indirect interaction of HuR with CRM1 through pp32 and APRIL, two proteins which contain CRM1-binding NES domains; the binding of pp32 and APRIL to HuR was mapped to RRM3.[27] In a subsequent study, Transportin-2 (Trn2) was identified as nuclear export receptor for HNS.[28] Later, Güttinger et al. demonstrated that Trn2 also acts as a nuclear import receptor for HuR in a Ran-GTPase dependent manner.[29] Independent findings by the Steitz lab showed that both, Trn1 and Trn2 can function as redundant import receptors for HuR.[30] Interestingly, stress induced by inhibition of RNA Polymerase II (pol II) transcription by actinomycin D resulted in almost quantitative relocalization of HuR, as well as HNS-containing reporters to the cytoplasm,[26] showing that HNS is sufficient to promote both nuclear localization in quiescent cells and stress induced export. What remains unclear is whether any observed net relocalization of HuR to the cytoplasm is due to an actual stimulation of HNS-mediated export or an indirect consequence of inhibited re-import into the nucleus.

REGULATION OF THE REGULATOR

In vitro and in vivo studies have shown that HuR is involved in the regulation of a large number of transiently expressed genes with functions as diverse as cell cycle regulators, growth factors, signal transducers, hormones, enzymes, metabolic factors, cytokines, chemokines, or cell surface receptors (ref. 19 and reviewed e.g., in refs. 8,31,32). HuR is generally considered a ubiquitously expressed protein; hence, the obvious question is how such a master-regulator is itself being regulated.

Different lines of evidence indicate that binding of HuR to the mRNA may be sufficient to promote all the regulatory downstream events, including a response at the protein level. Artificial recruitment of HuR to IL-2, IL-1 or TNF-α mRNAs by "mRNA openers", small antisense RNAs designed to render a cognate NNUUNNUUU motif accessible, was sufficient to promote selective, HuR-dependent stabilization of the targeted mRNA in cell-free systems.[19] In numerous studies, HuR binding to mRNA

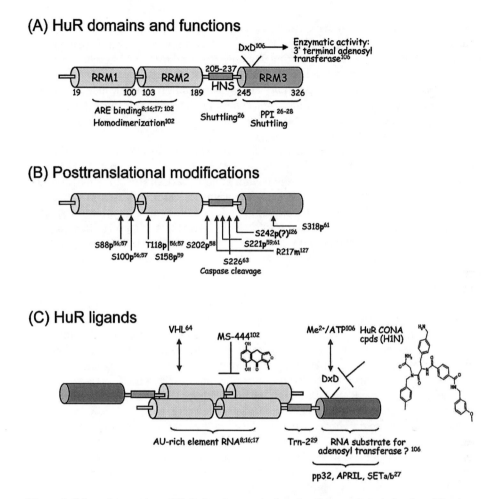

Figure 1. Schematic overview of HuR domain organization, functions, posttranslational modifications and ligands (adapted from ref. 106 and ref. 47). A) HuR domain organization and functional annotation. B) Positions of posttranslational modifications of HuR. Phosphorylation of S88, S100 as well as T118 is catalyzed by Chk2.[56,57] S158 and S221 are phosphorylated by PKCa.[59] S221 can also be phosphorylated by PKCd, which also targets S318.[61] S202 is a substrate for Cdk1.[58] Mutation of S242 to aspartate to mimic phosphorylation, but not mutation to alanine, led to retention of HuR in the nucleus.[126] However, direct evidence that this amino acid is indeed a target for phosphorylation remains to be shown. Arg217 is methylated by CARM-1.[127] Finally, HuR can be cleaved by caspase3 at S226. C) Overview of currently known RNA, protein and small molecule ligands of HuR. HuR is functional as a high-affinity homodimer, mediated via a dimerization interface within RRM1/2.[102] Binding to the ARE RNA has been assigned to these tandem RRMs,[8,16,17] with each HuR monomer contacting a single stranded NNUUNNUUU nonamer.[102] An interaction with protein ligands has been described for RRM1 (Von-Hippel Lindau tumor suppressor protein, VHL[64]) as well as for the hinge region and RRM3 (pp32, APRIL, SETα/β[128]). Small molecule HuR inhibitors which have been described to date comprise RRM1/2 targeted inhibitors of HuR homodimerization (MS-444, Dehydromutactin, Okicenone[102]) as well as ligands interacting with the ATP binding pocket in RRM3, such as the small molecule H1N1.[106] Flavonoids, which were recently described to inhibit HuR and HuC[129,130] are close analogues of MS-444 and therefore likely target a related binding pocket within RRM1/2.

was found to coincide with stabilization and/or increased expression of the encoded gene. Thus, a tight control of access of HuR to the target mRNA may be essential for keeping the system in check. Considering that both HuR and many of its target mRNAs are generally expressed in the same cell at the same time, there are generally two main possibilities for ensuring target specificity and dynamic on-off control: modulation of the HuR-RNA recognition by modifications acting on either the mRNA or the protein partner (including conformational changes), or control of their subcellular localization, or a combination of these mechanisms.

The movement of HuR from the nucleus to the cytoplasm has been correlated with its ability to stabilize ARE-containing mRNAs.[33,34] Following induction of MAPK via Anisomycin, for example, the 30 min time required for the exit of HuR from the nucleus coincided with stabilization of the β2-adrenergic receptor transcript.[35,36] Based on numerous studies it is now well established that export of HuR into the cytoplasm is a prerequisite for its protective effects on cognate target mRNAs.[25,37] Also for the neuronal Hu proteins, cytoplasmic localization is required for their function and unlike HuR, HuD as well as HuB are generally localized in the cytoplasm. HuD mutants lacking the nuclear export signal dominantly inhibited the function of the wild-type protein in promoting neurite outgrowth in PC12 neuroblasts.[38] Interestingly, aberrant and constitutive cytoplasmic localization of HuR was observed in a number of cancers and found to correlate with upregulation of proliferative, anti-apoptotic, pro-angiogenic and pro-metastatic genes and to be prognostic for poor patient outcome.[39-46] While this has raised interest in HuR as a potential drug target for cancer therapy, the question remains as to whether overexpression and cytoplasmic localization of HuR is a cause or consequence of the disease.

To unravel the mechanisms underlying dynamic control of HuR trafficking, extensive effort has been made to identify upstream signaling pathways as well as posttranslational modifications of HuR itself that modulate its localization (Fig. 1C).[47] Physiologically, HuR nucleocytoplasmic transport can be induced by many different stimuli, including specific triggers such as growth factors, activation of immune cells, or particular receptors. Relocalization into the cytoplasm is also observed in response to different kinds of stress, i.e., UV irradiation,[48,49] oxidative stress,[50] starvation,[51] or global transcription block.[26] These triggers act on HuR likely through a limited number of kinase-controlled signaling cascades, including the mitogen-activated protein kinases (MAPKs) and their upstream kinase MK-2,[52-54] the AMP-activated kinase (AMPK),[55] the cell-cycle checkpoint kinase 2 (Chk2),[56,57] cyclin dependent kinase 1 (Cdk1),[58] and members of the protein kinase C (PKC) family.[59-61] A direct posttranslational modification of HuR in response to these signaling events was found for Chk2, Cdk1, PKCα and PKCδ, as well as for the coactivator-associated arginine methyltransferase 1 (CARM1[62]), summarized in Figure 1B. Most of these modifications were implicated in mediating either HuR nuclear import or export, as comprehensively reviewed in reference 47. Interestingly, specific phosphorylation at S202 by Cdk1 results in sequestration of HuR by 14-3-3 proteins— an elegant mechanism to keep HuR in check during apoptosis.[58] Of note, cleavage of cytoplasmic HuR at S226 by caspase 3 in response to lethal stress was found to convert its role from an anti-apoptotic into a pro-apoptotic factor,[63] although the molecular details of its function remain to be revealed.

Besides influencing cellular trafficking, there are also indications of a direct impact of HuR phosphorylations events on RNA-binding. Among these are phosphorylation of S88, T118 and S100 which are all catalyzed by the cell cycle checkpoint kinase Chk2.[57] Abdelmohsen et al. suggested that HuR phosphorylation at S100 may reduce

RNA-binding, while phosphorylation at S88 and T118 (located in RRM2) may result in increased RNA-binding.[57] Based on our crystal structure of HuR RRM1 (PDB entry 3HI9, ref. 131) and in superposition with the crystal structure of RRM1 and RRM2 of HuD in complex with the c-fos ARE (PDB entry 1FXL, ref. 18), one may speculate about the structural impact of these modifications: It seems plausible that phosophorylation at HuR S100 may influence RNA-binding by inducing changes in RRM2-interdomain linker interactions. A phosphate at S88 would be expected to come in molecular proximity to the RNA, which, however, is more likely to repel the phosphate backbone.

In addition to posttranslational modifications, also protein ligands of HuR were reported to modulate its binding to target mRNAs. Datta et al. reported that binding of von Hippel Lindau (VHL) tumor suppressor protein via its elongin-binding domain to HuR RRM1 may compete with binding of HuR to VEGF mRNA, resulting in a shortened mRNA half-life and downregulation of VEGF expression.[64] More recently, a direct protein-protein interaction with the ARE-binding protein RNPC1 was reported to cooperatively increase the affinity of HuR to the p21 mRNA.[65]

Despite the general assumption that HuR is a ubiquitously expressed protein, there *are* some indications of dynamic changes of its levels. For example, nitric oxide was described to accelerate MMP9 mRNA decay by downregulating HuR expression.[66] As shown in a series of more recent studies, HuR expression may be subject to both transcriptional and post-transcriptional control. Based on HuR expression studies in renal tubular cells, Jeyaraj et al. suggested a model for HuR transcriptional and translational control by Smad proteins after renal injury.[67] Under basal conditions, two isoforms of the HuR transcript with different 5'UTRs were observed. In contrast to a short ~20 nt 5'UTR isoform, the longer ~150 nt 5'UTR isoform was poorly translatable. In response to bone morphogenetic protein 7 (Bmp7), Smad1 was found to bind to an alternative Smad1/5/7 binding site within the genomic region corresponding to the 150 nt 5'UTR and shift transcription towards production of the shorter, translatable isoform, thereby promoting an increase in the HuR protein level. Also the 3'UTR of HuR mRNA was found to be subject to various post-transcriptional control mechanisms. Interestingly, HuR was found to bind to an ARE in its own mRNA, which may result in an auto-regulatory feedback loop.[68] Again, different transcript isoforms were identified with varying lengths of the 3'UTR, generated by alternative polyadenylation. The distal portion of the 3'UTR, containing the ARE, was lost in the shorter, more abundant isoform. This part of the 3'UTR rendered reporter mRNAs unstable, but binding of HuR to it resulted in mRNA upregulation due to the decreased mRNA turnover.[68] More recently, additional HuR-mediated feedback mechanisms were reported to promote nucleocytoplasmic export of HuR mRNA.[69] However, it is to be expected that mechanisms are also in place to counterbalance this positive feedback loop to prevent the system to turn into a self-accelerating mode—a situation which may be dangerous for the organism. In fact, HuR overexpression is associated with a number of different cancers[39-45,70] and even if it is not yet clear whether this is sufficient to cause malignancies, increased HuR expression emerges at least as a significant contributing factor for tumor onset as well as progression. Interestingly, the anti-metastatic effects of green tea catechins, targeting the laminin receptor, were recently reported to be, at least in part, due to a downregulation of HuR and, subsequently, post-transcriptional reduction of the matrix metalloproteinase 9 (MMP9) expression.[71] A peptidic compound targeting the same receptor has recently gone through Phase IIa clinical trials in patients with prostate cancer refractory to hormone therapy (PCK3145, Ambrilia Biopharma). Finally, there is also evidence for miRNA-mediated control of HuR expression. miR-519 was described as the first miRNA to target HuR mRNA.[72] Based on experiments involving overexpression of miR-519 or its sequestration by anti-miRs, Abdelmohsen et al. showed that miR-519 represses HuR translation by targeting two sites, one in the coding region and one in the 3′UTR. More recently, it was shown that also miR-125a can translationally repress HuR.[73] Overexpression of the miR-125 precursor in MCF-7 breast cancer cell lines resulted in decreased proliferation and promotion of apoptosis, which was partly rescued by HuR re-expression. Given all these possible mechanisms through which the master-regulator HuR can itself be regulated, it will be interesting to further explore the defects that are responsible for HuR overexpression in malignancies and inflammatory diseases.

MOLECULAR MECHANISMS OF POST-TRANSCRIPTIONAL CONTROL BY HuR

While it is now well established *that* HuR is an essential factor modulating expression of a large variety of ARE controlled genes, we still face a lack of understanding in the underlying molecular mechanisms. For numerous targets, it has been shown that HuR interferes with ARE-dependent rapid mRNA turnover (reviewed e.g., in refs. 8,74). However, the molecular mechanism of this effect is still not clear. Initial reports of direct binding of Hu proteins via their RRM3 to polyA-RNA, as supported by crosslinking experiments,[75] would have suggested an elegant and plausible mechanism involving direct binding and protection of the mRNA polyA tail by ARE-bound HuR. However, evidence for such a mechanism is still missing. In fact, overexpression experiments suggested that HuR primarily protects the body of mRNA from degradation rather than slowing down mRNA deadenylation.[74] Additionally, HuR was shown to also modulate translation of a number of target mRNAs. In most of these cases, it was found to re-activate translation (e.g., p53,[76-78] ProTa,[79] HIF1a,[80] Cytochrome c,[81] MKP-1,[82] GLUT1,[83] CAT-1[84]) however, there are also a few examples in which HuR *promotes* ARE-mediated translational repression (e.g., c-Myc,[56,85] p27,[86,87] Wnt5a).[88] Interestingly, an additional role as control factor in alternative polyadenylation was described for nuclear HuR.[89] Finally, it is not surprising to find that viruses have found ways to hijack such a general host pathway for promoting gene expression. Hepatitis C Virus (HCV) as well as Human Immunodeficiency Virus HIV-1 were found to exploit HuR as host factor not only for nuclear export and stabilization of their RNAs, but also for RNA replication[90,91] and internal ribosome entry site (IRES)-mediated translational activation.[92]

Notably, the molecular mechanisms underlying these diverse functions of HuR as post-transcriptional regulator are still not understood. While it is generally assumed that HuR stabilizes mRNAs primarily by competing with decay factors for ARE binding, some data in the literature suggest that other mechanisms are also involved. For example, while HuR was found to compete in vitro for ARE binding with one of its main antagonists, AUF-1, in intact cells the two proteins were found concurrently bound on the same message.[93] Follow-up studies using fluorescence resonance energy transfer (FRET) in live or fixed cells then revealed that HuR and AUF1 are in sufficient proximity to associate with each other in both the nucleus and the cytoplasm.[36] Further interactions were reported for HuR with the RRM protein RNPC1 and the two proteins were found to cooperatively act on p21 3′UTR to result in mRNA stabilization.[65] Similarly, crosstalk of HuR on the Cox-2 3′UTR with the RNA-binding proteins CUGBP2[94] or RBM3[95] were found to modulate Cox-2 translation.

One important level for HuR to control the fate of mRNAs might be through modulating their subcellular localization. Besides its ability to transport mRNAs into the cytoplasm, HuR was also found to move between different RNP complexes and/or compartments within the cytoplasm. Consistent with its role in translational activation, for example, HuR was found to transiently associate, together with its cognate transcripts, with ribosomes and large polysomes in response to different stimuli.[96,97] Under various conditions of stress, however, not only relocalization of the protein into the cytoplasm is observed but also association with stress granules.[36,98] These cytoplasmic foci represent an accumulation of translationally arrested mRNAs and putatively function as reversible storage compartment of mRNAs which are temporarily not required during the stress response.[99,100] The presence of HuR in stress granules is therefore plausible, as its role may be not only to protect mRNAs from degradation during stress but also to rapidly shuttle them into polysomes and promote translation once the stress is relieved. In germ cells, dynamic localization of HuR in chromatoid bodies, sites of RNA storage and processing, has also been noted.[97] Altogether, this suggests that HuR helps controlling the fate of its target mRNAs by promoting their localization into sites of translation, translational arrest or mRNA turnover in dependence of cellular conditions. While plausible, direct evidence that HuR is an active transport factor rather than being dragged along with the associated mRNA remains to be provided.

Another interesting mechanistic aspect is that HuR itself seems to function as a homodimer. A self-association of Hu proteins was noted early on in yeast two-hybrid screens[101] and a band migrating at the size of the dimer on nonreducing SDS-PAGE was observed for recombinant HuR in several studies (e.g., ref. 102). FRET experiments then confirmed the presence of HuR homodimers in both the nucleus and the cytoplasm of live as well as fixed cells.[36] Using a chemical biology approach, the homodimerization was finally quantitatively characterized in vitro and revealed a self-association with subnanomolar affinity, mediated by RRM1 and RRM2; each monomer in the dimer bound one NNUUNNUUU containing RNA fragment.[102] High affinity low molecular weight inhibitors of HuR dimerization, identified in the same study, showed that only the dimer is capable of ARE binding in vitro. These compounds also inhibited HuR relocalization into the cytoplasm upon T-cell activation[102] as well as in response to several other triggers (Meisner et al., unpublished data). These data suggest that HuR either needs to be present as homodimer, or be bound to the mRNA to pass through the nuclear pore. The crystal structure of HuR RRM1, solved recently by our group, identifies an extensive hydrophobic interface between two RRM1 monomers at the α-helical backbone of the canonical RRM fold. However, an intermolecular disulfide bridge linking the cysteins located in the flexible N-terminal domain of the two monomers seems to play an essential role for at least nucleating the dimer formation (PDB entry 3HI9, Benoit et al., unpublished). In addition to RNA-independent homodimerization, an RNA-dependent oligomerization was reported for Drosophila ELAV,[103,104] HuD[101] as well as HuR.[105] Most plausibly, this may involve a stepwise and cooperative association of HuR monomers or preformed HuR dimers with mRNA, following nucleation of the interaction at the ARE. Finally, the recently discovered terminal transferase activity of HuR RRM3 opens still another mechanistic perspective.[106] A scenario where HuR plays an active role in RNA metabolism by modifying RNA 3'termini appears intriguing. However, physiological RNA substrates and functional significance of the RNA modification by HuR remain to be revealed.

Altogether, these studies suggest that HuR does not only act by displacing its antagonists from the ARE, but that is also involved in a more complex network of interactions with itself and other RBPs, and also miRNPs, on cellular mRNAs.

FUNCTION OF HuR IN THE RELIEF OF miRNA-MEDIATED REPRESSION

In the light of the aforementioned evidence that HuR has, generally, a positive affect on mRNA accumulation and sometimes also upregulates translation of target mRNAs (see above), we hypothesized that these effects of HuR might, at least in some situations, be due to this protein alleviating the inhibitory function of miRNAs. Similarly to HuR and many other regulatory proteins, miRNAs interact with mRNA 3'UTRs. However, their effect is opposite to that of HuR: miRNAs inhibit protein synthesis by repressing translation and/or destabilizing mRNAs (reviewed by ref. 107). In cells grown under normal conditions, HuR is primarily localized in the nucleus, but upon subjecting cells to different types of stress it relocates to the cytoplasm (see above). Hence, it was plausible to investigate the effect of stress on the miRNA-mediated repression and a potential regulatory role of HuR in this process. A more general objective of the study was to find out whether the miRNA-mediated inhibition is a regulated process and whether the repression can be reversed under specific cellular conditions.

miR-122 and liver cells grown in culture were chosen as a system most suitable to investigate the effect of HuR on miRNA-mediated repression. miR-122 represents the most abundant known tissue-specific miRNA. It is expressed almost exclusively in liver hepatocytes at over 50,000 copies per cell.[108,109] miR-122 is also expressed in human hepatoma Huh7 cells grown in culture but not in cultured hepatoma HepG2 cells, which can then serve as a convenient control. CAT-1 mRNA was selected as a model mRNA to study the regulation. This mRNA encodes the high-affinity cationic amino acid transporter, CAT-1, which facilitates uptake of arginine and lysine in mammalian cells. CAT-1 protein is expressed ubiquitously but its expression is known to undergo extensive regulation at transcriptional and post-transcriptional levels (reviewed by ref. 110). Tight regulation of CAT-1 activity and control of the uptake of arginine are particularly important in liver cells to avoid hydrolysis of the plasma arginine by arginase, a highly expressed enzyme in hepatocytes. Importantly, the CAT-1 mRNA is subject to regulation by both miR-122 and HuR. Several miR-122 sites are present in the 3'UTR of mouse and human CAT-1 mRNAs (Fig. 2A) and a role of miR-122 in regulation of CAT-1 synthesis has been demonstrated for both mRNAs.[84,109] Regulation of CAT-1 mRNA by HuR was previously established in rat glioma cells.[51]

Different types of stress (i.e., amino acid starvation, endoplasmic reticulum (ER) or oxidative stress), applied to Huh7 cells, were found to increase rapidly, within one hour, expression of either endogenous CAT-1 or luciferase reporters bearing the CAT-1 3'UTR. The induction occurred without an appreciable change in mRNA levels and was independent of Pol II transcription, since neither actinomycin D nor α-amanitin, inhibitors of pol II, had any effect. In contrast, the stimulation was inhibited by cycloheximide, an inhibitor of translational elongation. Taken together, these data indicated that the stimulation is caused by a translational mobilization of the pre-existing CAT-1 mRNA.[84]

The luciferase reporters bearing different fragments of the CAT-1 mRNA 3'UTR were then used to dissect the requirements for the stress-mediated activation. The induction of

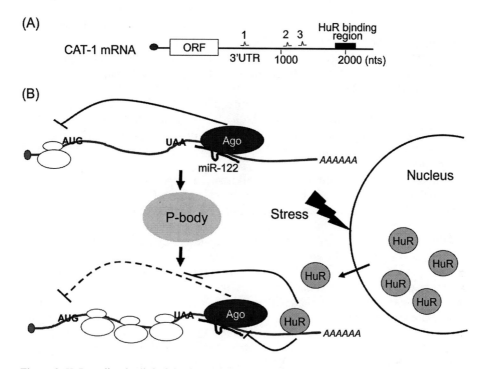

Figure 2. HuR-mediated relief of the CAT-1 mRNA repression by miR-122. A) Scheme of the human CAT-1 mRNA, with positions of the three miR-122 sites and a region interacting with HuR indicated. B) Summary of events occurring during the stress-mediated relief of repression. The repressed mRNA, shown in the upper part, is localized in P bodies. Upon stress, HuR translocates from the nucleus to the cytoplasm, binds to the CAT-1 3'UTR in a region positioned downstream of miR-122 sites and causes the relief of repression by either displacing the miRNP from mRNA or preventing its inhibitory function. Binding of HuR to the CAT-1 3'UTR also leads to the exit of the mRNA from P bodies and its recruitment to polysomes (for details, see text and ref. 84).

luciferase activity was only observed with reporters bearing both the miR-122 and HuR sites. Deletion of the HuR binding sites, positioned in the downstream portion of the CAT-1 3'UTR, completely eliminated the stimulatory effect of stress on translation, but the effect could be rescued by insertion of heterologous HuR sites, originating from other mRNAs. As discussed earlier in this chapter, HuR translocates from the nucleus to the cytoplasm in response to different forms of stress. RNA interference (RNAi)-mediated knock-down experiments were performed in order confirm that HuR is indeed essential for the activation of translation in the cytoplasm in response to stress. Depletion of HuR with either of two different small interfering RNAs (siRNAs) resulted in elimination of the stress effect. In addition, immunoprecipitation experiments showed that HuR specifically associates with CAT-1 mRNA in the cytosolic fraction from cells subjected to the amino acid starvation stress but not in unstressed control cells.[84]

The stimulatory effect of stress on CAT-1 mRNA or luciferase/CAT1 reporters was strictly dependent on the presence of miR-122 sites in the 3'UTR and the availability of miR-122. Consistently, stressing of hepatoma HepG2 cells, which do not express miR-122 and in which synthesis of CAT-1 is not repressed by miRNA, had no stimulatory effect on CAT-1 or reporter mRNA translation. Similarly, stress had no stimulatory effect on CAT-1 mRNA translation in Huh7 cells when activity of miR-122 was blocked by transfecting the cells with anti-sense oligonucleotides (anti-miRs) complementary to miR-122. Importantly, the effect of stress and HuR was not limited to mRNAs repressed by miR-122 but also occurred with reporters containing sites recognized by another miRNA. Activity of mRNA bearing sites specific for the let-7 miRNA was upregulated by stress in HeLa cell, which abundantly express this miRNA. Again, the derepression was dependent on a combined presence of both the let-7- and HuR- binding sites in the 3'UTR. The latter findings indicated that the ability of HuR to suppress the miRNA repression is of general character and does not apply to just one specific miRNA or one type of cells.[84]

Immunofluoresence and in situ hybridization experiments provided some understanding of how HuR is bringing about the relief of miRNA inhibition. In nonstressed Huh7 cells, a considerable fraction of CAT-1 mRNA localizes to processing bodies (P bodies; also known as GW bodies), cytoplasmic structures implicated in translational repression and mRNA degradation (reviewed by refs. 111,112). P bodies were also previously linked with miRNA repression: they are enriched in miRNAs and miRNA-inhibited mRNAs.[113,114] In addition, Argonaute (AGO) and GW182 proteins, two key components of miRNA ribonucleoproteins (miRNPs), that acting as the effectors in miRNA repression, are concentrated in P body granules (reviewed by ref. 107). P bodies are devoid of ribosomes and most translational initiation factors, consistent with the idea that they represent aggregates of inactive mRNAs.[111,112] Upon stressing Huh7 cells, the CAT-1 mRNA was found to relocalize from P bodies to the soluble fraction of the cytosol. As indicated by the results of RNAi depletion experiments, this relocalization was dependent on HuR. In addition, this redistribution was associated with the increase in the fraction of CAT-1 mRNA bound to polysomes, which is diagnostic of enhanced mRNA translation. Notably, derepression of CAT-1 induced by transfection of the anti-miR-122 oligonucleotide was likewise accompanied by relocalization of CAT-1 mRNA out of P bodies and its increased association with polysomes, consistent with the miRNA inhibition acting at the translation initiation step. In HepG2 cells, which do not express miR-122, the CAT-1 mRNA is actively translated even in the absence of stress and is not concentrated in P bodies. However, in HepG2 cells transfected with miR-122, the CAT-1 mRNA became enriched in P bodies, supporting the idea that the repression and P body localization of CAT-1 mRNA are controlled by miR-122.[84]

The experiments presented above provided a first example for a cross-talk between miRNPs and RBPs interacting with a 3'UTR. They demonstrated that CAT-1 mRNA and reporters bearing its 3'UTR can be relieved from miR-122-mediated repression in human Huh7 hepatoma cells when they are subjected to different types of stress. They also showed that the derepression is accompanied by the release of CAT-1 mRNA from P bodies and its entry into polysomes and that the process involves binding of HuR to ARE-like elements in the 3'UTR of CAT-1 mRNA (illustrated in Fig. 2). The demonstration that mRNAs repressed by miRNAs can be mobilized from P bodies to return to active translation indicated that P bodies are dynamic structures, exchanging their content rapidly with that of the cytosol, as also suggested previously by photobleaching experiments (ref. 115, reviewed in refs. 111,112). Together with the studies analyzing the response of miRNA regulation to synaptic stimulation in neurons[116,117] and additional examples of effects of RBPs on miRNA repression[118,119] the CAT-1/miR-122/HuR findings indicated that miRNA regulation is much more dynamic than previously anticipated and is able to respond rapidly to specific cellular needs.

How could HuR interaction with the mRNA 3'UTR lead to reversal of the inhibitory effect of miRNAs on translation? It is unlikely that this is due to the protein competing with miRNAs for the same or overlapping binding sites on a target mRNA. In CAT-1 mRNA and also some of the tested reporters, HuR binding sites are positioned up to several hundreds nucleotides away from sequences recognized by miRNAs. However, it is possible that HuR, even when positioned at the distance, is brought in proximity of miRNP by RNA folding, which in turn might facilitate interaction of the protein with some miRNP components and either lead to the dissociation of miRNPs from the mRNA or prevent them from acting as effectors in the repression (Fig. 2B). Alternatively, interaction of HuR at sequences positioned even far away from miRNA sites could nucleate oligomerization of the protein and its "spreading" along the RNA sequence, leading to the displacement of miRNP from mRNA. Indeed, RNA-dependent formation of oligomers of HuR and some other ELAV family proteins has been previously described[104,105] (see also above). Another interesting possibility is raised by recent findings that HuR has enzymatic activity, catalyzing addition of adenylate residues to the RNA 3' end.[106] Could HuR affect activity or stability of an mRNP by inducing modification of its miRNA component?

It is also not clear how HuR induces the relocation of translationally repressed mRNAs from P bodies. As already discussed above, P bodies are dynamic structures, exchanging their content rapidly with that of the cytosol. Thus, it is possible that HuR, present in abundance in the cytoplasm of stressed cells, shifts the P-body-to-cytosol equilibrium of repressed mRNAs by binding to their 3'UTRs. The transfer into the cytosol would then facilitate re-entry of mRNAs to active translation. This scenario takes into account the observation that HuR, even in cells subjected to different stress conditions, is not a component of P bodies.[84] Clearly, additional studies are required to establish molecular understanding of the HuR role in both the relocation of repressed mRNAs from P bodies and their relief from the inhibitory function miRNAs.

SYNERGISM BETWEEN HuR AND let-7 IN TRANSLATIONAL REPRESSION OF c-Myc mRNA

In contrast to the situation with CAT-1 mRNA, repression of mRNA encoding the proto-oncogene c-Myc by the let-7 miRNA is enhanced by the binding of HuR to AREs adjacent to the let-7 site.[120] The c-Myc mRNA was identified as a target of HuR already long time ago,[85] but Kim et al.[120] investigated the relevance of this interaction in more detail. In initial experiments they found that, in HeLa cells, HuR has repressive effect on the expression of c-Myc at both mRNA and protein levels and that this effect is mediated by the 3'UTR. Reporters bearing the c-Myc 3'UTR were subjects to similar regulation and pull-down analysis confirmed the binding of HuR to the 3'UTR. Kim et al. then identified a site in the c-Myc 3'UTR predicted to interact with let-7b or let-7c miRNAs and confirmed that these miRNAs indeed exert repressive effects on the level of either endogenous c-Myc mRNA and protein or reporters bearing the c-Myc 3'UTR. Surprisingly, they found that repression by let-7 was completely dependent on the binding of HuR to the region adjacent to the miRNA site. Consistently, depleting cells of HuR abrogated let-7-mediated inhibition of c-Myc. It also diminished association of Ago-2 with the 3'UTR, supporting the involvement of miRNP in the regulation. Additional control experiments, involving use of reporters bearing mutations in the let-7 site, indicated that HuR itself has no repressive effect on expression of c-Myc; instead, it functions in promoting association of let-7 miRNP with the 3′UTR.

The requirement of HuR for the let-7-induced repression of c-Myc contrasts with the stress-induced suppression of the inhibitory effect of miRNAs on expression CAT-1 and reporter mRNAs investigated by Bhattacharyya et al.[84] It is difficult to explain these two different outcomes of the HuR regulation. In c-Myc 3′UTR, the let-7 and HuR sites are located relatively close to each other. Hence, it is possible that HuR binding modifies local RNA secondary structure to unmask the let-7 recognition site. Such scenario would be unlikely to operate in the case of CAT-1 mRNA, in which HuR and miR-122 sites are positioned very far apart.[84] It should be noted that while the CAT-1 mRNA regulation was studied in cells subjected to different stress conditions, the effects on c-Myc were investigated in nonstressed cells. Clearly, the cytoplasmic concentration of HuR, but possibly also a pattern of its posttranslational modifications, might differ between stress and nonstressed cells.

CONCLUSION

Results described in this and other chapters of the book indicate that regulation of miRNA repression by RBPs is probably a widespread phenomenon. In addition to HuR, also Dnd1[118] and APOBEC3G (apolipoprotein B mRNA-editing enzyme catalytic polypeptide-like 3G; ref. 119) were already shown to regulate miRNA repression in mammalian cells and/or zebrafish. Recent comparative studies of mRNAs interacting with Pumilio (PUF) proteins, which have been linked to let-7 repression of hbl-1 mRNA in *C. elegans*,[121] showed a considerable enrichment of PUF binding sites in the vicinity of predicted miRNA recognition sequences in human mRNAs,[122] suggesting a regulatory role also for PUF proteins. An important challenge will be to elucidate mechanisms underlying the function of all the aforementioned proteins in regulating miRNA repression. For HuR, it will be essential to investigate a potential role of posttranslational modifications of this protein and its catalytic activity to adenylate the RNA 3′ end. It will be important to identify new RBPs participating in the regulation of miRNA function. For example, given that many miRNAs are specifically expressed in the brain[123] and that three of the four mammalian ELAV proteins, HuB, HuC and HuD, are restricted to neurons,[13,14] it is possible that these proteins play a regulatory role in miRNA-mediated repression in neuronal cells. In neurons, many mRNAs are transported along the dendrites as repressed mRNPs to become translated at dendritic spines upon synaptic activation. miRNAs have already been implicated in reversible control of translation at synapses (reviewed by ref. 123) and it will be important to investigate contribution of RBPs to this type of regulation. New, high throughput technologies, such as HITS-CLIP (ref. 24), will greatly facilitate identification of mRNA sites recognized by different RBPs in vivo and in establishing relationships between RBPs and miRNPs.

Vasudevan et al.[124,125] have recently reported that, in serum-starved or nonproliferating human cells, the AGO2-miRNA complex bound to the 3′UTR of TNFα mRNA recruits the Fragile-X-Related Protein 1 (FXR1; a well characterized RBP), which results in miRNA-dependent stimulation, rather then inhibition, of translation of target mRNA. This finding revealed still one additional possible effect of the interplay between miRNPs and RBPs interacting with mRNA 3′UTRs. It is probably safe to conclude that different combinations of RBPs and miRNAs will have very different effects on the outcome of translation, strongly dependent on the identity of the target mRNA. The spectrum of the effects will extend from the increase in translational repression, through its relief, to translational activation. Similar broad range of effects will likely apply to modulation of mRNA stability. Although, thus far, these were generally effects of RBPs on the activity of miRNPs that have been scrutinized, no doubt also miRNPs will be found in the future to modulate function of specific RBPs involved in post-transcriptional control.

ACKNOWLEDGEMENTS

The authors thank David V. Morrissey for proof-reading of the manuscript. The work done in the laboratory of WF is supported by the EC FP6 Program "Sirocco". The Friedrich Miescher Institute is supported by the Novartis Research Foundation.

REFERENCES

1. Chen CY, Shyu AB. Selective degradation of early-response-gene mRNAs: functional analyses of sequence features of the AU-rich elements. Mol Cell Biol 1994; 14:8471-8482.
2. Barreau C, Paillard L, Osborne HB. AU-rich elements and associated factors: are there unifying principles? Nucleic Acids Res 2006; 33:7138-7150.
3. Bolognani F, Perrone-Bizzozero NI. RNA-protein interactions and control of mRNA stability in neurons. J Neurosci Res 2008; 86:481-489.
4. Bickel M, Iwai Y, Pluznik DH et al. Binding of sequence-specific proteins to the adenosine- plus uridine-rich sequences of the murine granulocyte/macrophage colony-stimulating factor mRNA. Proc Natl Acad Sci USA 1992; 89:10001-10005.
5. Espel E. The role of the AU-rich elements of mRNAs in controlling translation. Semin. Cell Dev Biol 2005; 16:59-67.
6. Chen CY, Shyu AB. AU-rich elements: characterization and importance in mRNA degradation. Trends Biochem Sci 1995; 20:465-470.
7. Bakheet T, Williams BR, Khabar KS. ARED 3.0: the large and diverse AU-rich transcriptome. Nucleic Acids Res 2006; 34:D111-D114.
8. Brennan CM, Steitz JA. HuR and mRNA Stability. Cell Mol Life Sci 2001; 58:266-277.
9. Gingerich TJ, Feige JJ, LaMarre J. AU-rich elements and the control of gene expression through regulated mRNA stability. Anim Health Res Rev 2004; 5:49-63.
10. Blackshear PJ. Tristetraprolin and other CCCH tandem zinc-finger proteins in the regulation of mRNA turnover. Biochem Soc Trans 2002; 30:945-952.
11. Zhang T, Kruys V, Huez G et al. AU-rich element-mediated translational control: complexity and multiple activities of trans-activating factors. Biochem Soc Trans 2002; 30:952-958.
12. Garneau NL, Wilusz J, Wilusz CJ. The highways and byways of mRNA decay. Nat Rev Mol Cell Biol 2007; 8:113-126.
13. Perrone-Bizzozero N, Bolognani F. Role of HuD and other RNA-binding proteins in neural development and plasticity. J Neurosci Res 2002; 68:121-126.
14. Pascale A, Amadio M, Quattrone A. Defining a neuron: neuronal ELAV proteins. Cell Mol Life Sci 2008; 65:128-140.
15. Good PJ. A conserved family of elav-like genes in vertebrates. Proc Natl Acad Sci USA 1995; 92:4557-4561.
16. Ma W.-J, Cheng S, Campbell C et al. Cloning and Characterization of HuR, a Ubiquitously Expressed ELAV-like Protein. J Biol Chem 1996; 271:8144-8151.
17. Myer VE, Fan XC, Steitz JA. Identification of HuR as a protein imlicated in AUUUA-mediated mRNA decay. EMBO J 1997; 16:2130-2139.
18. Wang X, Tanaka Hall TM. Structural basis for recognition of AU-rich element RNA by the HuD protein. Nature Structural Biology 2001; 8:141-145.
19. Meisner NC, Hackermuller J, Uhl V et al. mRNA openers and closers: modulating AU-rich element-controlled mRNA stability by a molecular switch in mRNA secondary structure. Chembiochem 2004; 5:1432-1447.

20. De Silanes IL, Zhan M, Lal A et al. Identification of a target RNA motif for RNA-binding protein HuR. Proc Natl Acad Sci USA 2004; 101:2987-2992.
21. Yarovinsky TO, Butler NS, Monick MM et al. Early Exposure to IL-4 Stabilizes IL-4 mRNA in CD4+ T-Cells via RNA-Binding Protein HuR. J Immunol 2006; 177:4426-4435.
22. Winter J, Roepcke S, Krause S et al. Comparative 3'UTR analysis allows identification of regulatory clusters that drive Eph/ephrin expression in cancer cell lines. PLoS One 2008; 3:e2780.
23. Ule J, Jensen K, Mele A et al. CLIP: a method for identifying protein-RNA interaction sites in living cells. Methods 2005; 37:376-386.
24. Chi SW, Zang JB, Mele A et al. Argonaute HITS-CLIP decodes microRNA-mRNA interaction maps. Nature 2009; 460:479-486.
25. Fan XC, Steitz JA. Overexpression of HuR, a nuclear-cytoplasmic shuttling protein, increases the in vivo stability of ARE-containing mRNAs. EMBO J 1998; 17:3448-3460.
26. Fan XC, Steitz JA. HNS, a nuclear-cytoplasmic shuttling sequence in HuR. Proc Natl Acad Sci USA 1998; 95:15293-15298.
27. Gallouzi IE, Brennan CM, Steitz JA. Protein ligands mediate the CRM1-dependent export of HuR in response to heat shock. RNA 2001; 7:1348-1361.
28. Gallouzi IE, Steitz JA. Delineation of mRNA export pathways by the use of cell-permeable peptides. Science 2001; 294:1895-1901.
29. Guttinger S, Muhlhausser P, Koller-Eichhorn R et al. From The Cover: Transportin2 functions as importin and mediates nuclear import of HuR. Proc Natl Acad Sci USA 2004; 101:2918-2923.
30. Rebane A, Aab A, Steitz JA. Transportins 1 and 2 are redundant nuclear import factors for hnRNP A1 and HuR. RNA 2004; 10:590-599.
31. Cherry J, Karschner V, Jones H et al. HuR, an RNA-binding protein, involved in the control of cellular differentiation. In Vivo 2006; 20:17-23.
32. Abdelmohsen K, Lal A, Kim HH et al. Post-transcriptional orchestration of an anti-apoptotic program by HuR. Cell Cycle 2007; 6:1288-1292.
33. Atasoy U, Watson J, Patel D et al. ELAV protein HuA (HuR) can redistribute between nucleus and cytoplasm and is upregulated during serum stimulation and T-cell activation. J Cell Sci 1998; 111(Pt 21):3145-3156.
34. Xu YZ, Di MS, Gallouzi I et al. RNA-binding protein HuR is required for stabilization of SLC11A1 mRNA and SLC11A1 protein expression. Mol Cell Biol 2005; 25:8139-8149.
35. Headley VV, Tanveer R, Greene SM et al. Reciprocal regulation of beta-adrenergic receptor mRNA stability by mitogen activated protein kinase activation and inhibition. Mol Cell Biochem 2004; 258:109-119.
36. David PS, Tanveer R, Port JD. FRET-detectable interactions between the ARE binding proteins, HuR and p37AUF1. RNA 2007; 13:1453-1468.
37. Keene JD. Why is Hu where? Shuttling of early-response-gene messenger RNA subsets. Proc Natl Acad Sci USA 1999; 96:5-7.
38. Akamatsu W, Okano HJ, Osumi N et al. Mammalian ELAV-like neuronal RNA-binding proteins HuB and HuC promote neuronal development in both the central and the peripheral nervous systems. Proc Natl Acad Sci USA 1999; 96:9885-9890.
39. Dixon DA, Tolley ND, King PH et al. Altered expression of the mRNA stability factor HuR promotes cyclooxygenase-2 expression in colon cancer cells. J Clin Invest 2001; 108:1657-1665.
40. López de Silanes I, Fan J, Galbán CJ et al. Global analysis of HuR-regulated gene expression in colon cancer systems of reducing complexity. Gene Expr 2004; 12:49-59.
41. Erkinheimo TL, Sivula A, Lassus H et al. Cytoplasmic HuR expression correlates with epithelial cancer cell but not with stromal cell cyclooxygenase-2 expression in mucinous ovarian carcinoma. Gynecol Oncol 2005; 99:14-19.
42. Heinonen M, Fagerholm R, Aaltonen K et al. Prognostic role of HuR in hereditary breast cancer. Clin Cancer Res 2007; 13:6959-6963.
43. Hostetter C, Licata LA, Witkiewicz A et al. Cytoplasmic accumulation of the RNA binding protein HuR is central to tamoxifen resistance in estrogen receptor positive breast cancer cells. Cancer Biol Ther 2008; 7:1496-1506.
44. Hasegawa H, Kakuguchi W, Kuroshima T et al. HuR is exported to the cytoplasm in oral cancer cells in a different manner from that of normal cells. Br J Cancer 2009; 100:1943-1948.
45. Wang J, Zhao W, Guo Y et al. The expression of RNA-binding protein HuR in nonsmall cell lung cancer correlates with vascular endothelial growth factor-C expression and lymph node metastasis. Oncology 2009; 76:420-429.
46. López de Silanes I, Lal A, Gorospe M. HuR—Post-transcriptional Paths to Malignancy. RNA Biology 2005; 2:e11-e13.
47. Doller A, Pfeilschifter J, Eberhardt W. Signalling pathways regulating nucleo-cytoplasmic shuttling of the mRNA-binding protein HuR. Cell Signal 2008; 20:2165-2173.

48. Wang W, Furneaux H, Cheng H et al. HuR regulates p21 mRNA Stabilization by UV Light. Mol Cell Biol 2000; 20:760-769.
49. Westmark CJ, Bartleson VB, Malter JS. RhoB mRNA is stabilized by HuR after UV light. Oncogene 2005; 24:502-511.
50. Abdelmohsen K, Kuwano Y, Kim HH et al. Post-transcriptional gene regulation by RNA-binding proteins during oxidative stress: implications for cellular senescence. Biol Chem 2008; 389:243-255.
51. Yaman I, Fernandez J, Sarkar B et al. Nutritional control of mRNA stability is mediated by a conserved AU-rich element that binds the cytoplasmic shuttling protein HuR. J Biol Chem 2002; 277:41539-41546.
52. Winzen R, Kracht M, Ritter B et al. The p38 MAP kinase pathway signals for cytokine-induced mRNA stabilization via MAP kinase-activated protein kinase 2 and an AU-rich region-targeted mechansim. EMBO J 1999; 18:4969-4980.
53. Ming XF, Stoecklin G, Lu M et al. Parallel and independent regulation of interleukin-3 mRNA turnover by phosphatidylinositol 3-kinase and p38 mitogen-activated protein kinase. Mol Cell Biol 2001; 21:5778-5789.
54. Subbaramaiah K, Marmao TP, Dixon DA et al. Regulation of cyclooxygenase-2 mRNA stability by taxanes. Evidence for involvement of p38, MAPKAPK-2 and HuR. J Biol Chem 2003; 278:37637-37647.
55. Wang W, Fan J, Yang X et al. AMP-Activated Kinase Regulates Cytoplasmic HuR. Mol Cell Biol 2002; 22:3425-3436.
56. Liu L, Rao JN, Zou T et al. Polyamines regulate c-Myc translation through Chk2-dependent HuR phosphorylation. Mol Biol Cell 2009; 20:4885-4898.
57. Abdelmohsen K, Pullmann R, Jr., Lal A et al. Phosphorylation of HuR by Chk2 regulates SIRT1 expression. Mol Cell 2007; 25:543-557.
58. Kim HH, Abdelmohsen K, Lal A et al. Nuclear HuR accumulation through phosphorylation by Cdk1. Genes Dev 2008; 22:1804-1815.
59. Doller A, Huwiler A, Mueller R et al. Protein kinase C alpha-dependent phosphorylation of the mRNA-stabilizing factor HuR: implications for post-transcriptional regulation of cyclooxygenase-2. Mol Biol Cell 2008; 18:2137-2148.
60. Doller A, Akool e, Huwiler A et al. Posttranslational modification of the AU-rich element binding protein HuR by protein kinase Cdelta elicits angiotensin II-induced stabilization and nuclear export of cyclooxygenase 2 mRNA. Mol Cell Biol 2008; 28:2608-2625.
61. Doller A, Schlepckow K, Schwalbe H et al. Tandem phosphorylation of serine 221 and 318 by PKC{delta} coordinates mRNA binding and nucleo-cytoplasmic shuttling of HuR. Mol Cell Biol 2010; Epub.
62. Laird-Offringa IA, Elfferich P, van der Eb AJ. Rapid c-myc mRNA degradation does not require (A + U)-rich sequences or complete translation of the mRNA. Nucleic Acids Res 1991; 19:2387-2394.
63. Mazroui R, Di MS, Clair E et al. Caspase-mediated cleavage of HuR in the cytoplasm contributes to pp32/PHAP-I regulation of apoptosis. J Cell Biol 2008; 180:113-127.
64. Datta K, Mondal S, Sinha S et al. Role of elongin-binding domain of von hippel lindau gene product on HuR-mediated VPF/VEGF mRNA stability in renal cell carcinoma. Oncogene 2005; 24:7850-7858.
65. Cho SJ, Zhang J, Chen X. RNPC1 modulates the RNA-binding activity of and cooperates with, HuR to regulate p21 mRNA stability. Nucleic Acids Res 2010.
66. Akool e, Kleinert H, Hamada FM et al. Nitric oxide increases the decay of matrix metalloproteinase 9 mRNA by inhibiting the expression of mRNA-stabilizing factor HuR. Mol Cell Biol 2003; 23:4901-4916.
67. Jeyaraj SC, Singh M, Ayupova DA et al. Transcriptional control of human antigen R by bone morphogenetic protein. J Biol Chem 2010; 285:4432-4440.
68. Al-Ahmadi W, Al-Ghamdi M, Al-Haj L et al. Alternative polyadenylation variants of the RNA binding protein, HuR: abundance, role of AU-rich elements and auto-Regulation. Nucleic Acids Res 2009; 37:3612-3624.
69. Yi J, Chang N, Liu X et al. Reduced nuclear export of HuR mRNA by HuR is linked to the loss of HuR in replicative senescence. Nucleic Acids Res 2009; Epub.
70. Lopez DeSilanes I, Fan J, Yang X et al. Role of the RNA-binding protein HuR in colon carcinogenesis. Oncogene 2003; 22:7146-7154.
71. Annabi B, Currie JC, Moghrabi A et al. Inhibition of HuR and MMP-9 expression in macrophage-differentiated HL-60 myeloid leukemia cells by green tea polyphenol EGCg. Leuk Res 2007; 31:1277-1284.
72. Abdelmohsen K, Srikantan S, Kuwano Y et al. miR-519 reduces cell proliferation by lowering RNA-binding protein HuR levels. Proc Natl Acad Sci USA 2008; 105:20297-20302.
73. Guo X, Wu Y, Hartley RS. microRNA-125a represses cell growth by targeting HuR in breast cancer. RNA Biol 2009; 6:575-583.
74. Peng SS-Y, Chen C-YA, Xu N et al. RNA stabilization by the AU-rich element binding protein, HuR, an ELAV protein. EMBO J 1998; 17:3461-3470.
75. Ma W-J, Chung S, Furneaux H. The ELAV-like proteins bind to AU-rich elements and to the poly(A) tail of mRNA. Nucl Acids Res 1997; 25:3564-3569.

76. Mazan-Mamczarz K, Galban S, Lopez DS et al. RNA-binding protein HuR enhances p53 translation in response to ultraviolet light irradiation. Proc Natl Acad Sci USA 2003; 100:8354-8359.
77. Galban S, Martindale JL, Mazan-Mamczarz K et al. Influence of the RNA-binding protein HuR in pVHL-regulated p53 expression in renal carcinoma cells. Mol Cell Biol 2003; 23:7083-7095.
78. Tong X, Pelling JC. Enhancement of p53 expression in keratinocytes by the bioflavonoid apigenin is associated with RNA-binding protein HuR. Mol Carcinog 2009; 48:118-129.
79. Lal A, Kawai T, Yang X et al. Antiapoptotic function of RNA-binding protein HuR effected through prothymosin alpha. EMBO J 2005; 24:1852-1862.
80. Galban S, Kuwano Y, Pullmann R, Jr. et al. RNA-binding proteins HuR and PTB promote the translation of hypoxia-inducible factor 1alpha. Mol Cell Biol 2008; 28:93-107.
81. Kawai T, Lal A, Yang X et al. Translational control of cytochrome c by RNA-binding proteins TIA-1 and HuR. Mol Cell Biol 2006; 26:3295-3307.
82. Kuwano Y, Kim HH, Abdelmohsen K et al. MKP-1 mRNA stabilization and translational control by RNA-binding proteins HuR and NF90. Mol Cell Biol 2008; 28:4562-4575.
83. Gantt KR, Cherry J, Richardson M et al. The regulation of glucose transporter (GLUT1) expression by the RNA binding protein HuR. J Cell Biochem 2006; 99:565-574.
84. Bhattacharyya SN, Habermacher R, Martine U et al. Relief of microRNA-mediated translational repression in human cells subjected to stress. Cell 2006; 125:1111-1124.
85. Lafon I, Carballes F, Brewer G et al. Developmental expression of AUF1 and HuR, two c-myc mRNA binding proteins. Oncogene 1998; 16:3413-3421.
86. Millard SS, Vidal A, Markus M et al. A U-rich element in the 5' untranslated region is necessary for the translation of p27 mRNA. Mol Cell Biol 2000; 20:5947-5959.
87. Kullmann M, Gopfert U, Siewe B et al. ELAV/Hu proteins inhibit p27 translation via an IRES element in the p27 5'UTR. Genes Dev 2002; 16:3087-3099.
88. Leanderssn K, Riesbeck K, Andersson T. Wnt-5a mRNA translation is suppressed by the Elav-like protein HuR in human breast epithelial cells. Nucleic Acids Res 2006; 34:3988-3999.
89. Zhu H, Zhou HL, Hasman RA et al. Hu proteins regulate polyadenylation by blocking sites containing U-rich sequences. J Biol Chem 2007; 282:2203-2210.
90. Spångberg K, Wiklund L, Schwartz S. HuR, a Protein Implicated in Oncogene and Growth Factor mRNA Decay, Binds to the 3' Ends of Hepatitis C Virus RNA of Both Polarities. Virology 2000; 274:378-390.
91. Korf M, Jarczak D, Beger C et al. Inhibition of hepatitis C virus translation and subgenomic replication by siRNAs directed against highly conserved HCV sequence and cellular HCV cofactors. J Hepatol 2005; 43:225-234.
92. Rivas-Aravena A, Ramdohr P, Vallejos M et al. The Elav-like protein HuR exerts translational control of viral internal ribosome entry sites. Virology 2009; 392:178-185.
93. Lal A, Mazan-Mamczarz K, Kawai T et al. Concurrent versus individual binding of HuR and AUF1 to common labile target mRNAs. EMBO J 2004; 23:3092-3102.
94. Sureban SM, Murmu N, Rodriguez P et al. Functional antagonism between RNA binding proteins HuR and CUGBP2 determines the fate of COX-2 mRNA translation. Gastroenterology 2007; 132:1055-1065.
95. Sureban SM, Ramalingam S, Natarajan G et al. Translation regulatory factor RBM3 is a proto-oncogene that prevents mitotic catastrophe. Oncogene 2008; 27:4544-4556.
96. Blaxall BC, Pellett AC, Wu SC et al. Purification and characterization of beta-adrenergic receptor mRNA-binding proteins. J Biol Chem 2000; 275:4290-4297.
97. Nguyen CM, Chalmel F, Agius E et al. Temporally regulated traffic of HuR and its associated ARE-containing mRNAs from the chromatoid body to polysomes during mouse spermatogenesis. PLoS One 2009; 4:e4900.
98. Gallouzi I-E, Brennan CM, Stenberg MG et al. HuR binding to cytoplasmic mRNA is perturbed by heat shock. Proc Natl Acad Sci USA 2000; 97:3073-3078.
99. Kedersha N, Anderson P. Stress granules: sites of mRNA triage that regulate mRNA stability and translatability. Biochem Soc Trans 2002; 30:963-969.
100. Kedersha N, Stoecklin G, Ayodele M et al. Stress granules and processing bodies are dynamically linked sites of mRNP remodeling. J Cell Biol 2005; 169:871-884.
101. Kasashima K, Sakashita E, Saito K et al. Complex formation of the neuron-specific ELAV-like Hu RNA-binding proteins. Nucl Acids Res 2002; 30:4519-4526.
102. Meisner NC, Hintersteiner M, Mueller K et al. Identification and mechanistic characterization of low-molecular-weight inhibitors for HuR. Nat Chem Biol 2007; 3:508-515.
103. Soller M, White K. ELAV multimerizes on conserved AU4-6 motifs important for ewg splicing regulation. Mol Cell Biol 2005; 25:7580-7591.
104. Toba G, White K. The third RNA recognition motif of Drosophila ELAV protein has a role in multimerization. Nucleic Acids Res 2008; 36:1390-1399.

105. Fialcowitz-White EJ, Brewer BY, Ballin JD et al. Specific protein domains mediate cooperative assembly of HuR oligomers on AU-rich mRNA-destabilizing sequences. J Biol Chem 2007; 282:20948-20959.

106. Meisner NC, Hintersteiner M, Seifert JM et al. Terminal adenosyl transferase activity of post-transcriptional regulator HuR revealed by confocal on-bead screening. J Mol Biol 2009; 386:435-450.

107. Filipowicz W, Bhattacharyya SN, Sonenberg N. Mechanisms of post-transcriptional regulation by microRNAs: are the answers in sight? Nat Rev Genet 2008; 9:102-114.

108. Lagos-Quintana M, Rauhut R, Meyer J et al. New microRNAs from mouse and human. RNA 2003; 9:175-179.

109. Chang J, Nicolas E, Marks D et al. miR-122, a mammalian liver-specific microRNA, is processed from hcr mRNA and may downregulate the high affinity cationic amino acid transporter CAT-1. RNA Biol 2004; 1:106-113.

110. Hatzoglou M, Fernandez J, Yaman I et al. Regulation of cationic amino acid transport: the story of the CAT-1 transporter. Annu Rev Nutr 2004; 24:377-399.

111. Eulalio A, Behm-Ansmant I, Izaurralde E. P bodies: at the crossroads of post-transcriptional pathways. Nat Rev Mol Cell Biol 2007; 8:9-22.

112. Parker R, Sheth U. P bodies and the control of mRNA translation and degradation. Mol Cell 2007; 25:635-646.

113. Pillai RS, Bhattacharyya SN, Artus CG et al. Inhibition of translational initiation by Let-7 microRNA in human cells. Science 2005; 309:1573-1576.

114. Liu J, Valencia-Sanchez MA, Hannon GJ et al. microRNA-dependent localization of targeted mRNAs to mammalian P-bodies. Nat Cell Biol 2005; 7:719-723.

115. Andrei MA, Ingelfinger D, Heintzmann R et al. A role for eIF4E and eIF4E-transporter in targeting mRNPs to mammalian processing bodies. RNA 2005; 11:717-727.

116. Ashraf SI, McLoon AL, Sclarsic SM et al. Synaptic protein synthesis associated with memory is regulated by the RISC pathway in Drosophila. Cell 2006; 124:191-205.

117. Schratt GM, Tuebing F, Nigh EA et al. A brain-specific microRNA regulates dendritic spine development. Nature 2006; 439:283-289.

118. Kedde M, Strasser MJ, Boldajipour B et al. RNA-binding protein Dnd1 inhibits microRNA access to target mRNA. Cell 2007; 131:1273-1286.

119. Huang J, Liang Z, Yang B et al. Derepression of microRNA-mediated protein translation inhibition by apolipoprotein B mRNA-editing enzyme catalytic polypeptide-like 3G (APOBEC3G) and its family members. J Biol Chem 2007; 282:33632-33640.

120. Kim HH, Kuwano Y, Srikantan S et al. HuR recruits let-7/RISC to repress c-Myc expression. Genes Dev 2009; 23:1743-1748.

121. Nolde MJ, Saka N, Reinert KL et al. The Caenorhabditis elegans pumilio homolog, puf-9, is required for the 3'UTR-mediated repression of the let-7 microRNA target gene, hbl-1. Dev Biol 2007; 305:551-563.

122. Galgano A, Forrer M, Jaskiewicz L et al. Comparative analysis of mRNA targets for human PUF-family proteins suggests extensive interaction with the miRNA regulatory system. PLoS One 2008; 3:e3164.

123. Schratt G. microRNAs at the synapse. Nat Rev Neurosci 2009; 10:842-849.

124. Vasudevan S, Steitz JA. AU-Rich-Element-Mediated Upregulation of Translation by FXR1 and Argonaute 2. Cell 2007; 128:1105-1118.

125. Vasudevan S, Tong Y, Steitz JA. Switching from repression to activation: microRNAs can up-regulate translation. Science 2007; 318:1931-1934.

126. Kim HH, Yang X, Kuwano Y et al. Modification at HuR(S242) alters HuR localization and proliferative influence. Cell Cycle 2008; 7:3371-3377.

127. Li H, Park S, Kilburn B et al. Lipopolysaccharide-induced methylation of HuR, an mRNA-stabilizing protein, by CARM1. J Biol Chem 2002; 277:44623-44630.

128. Brennan CM, Gallouzi I-E, Steitz JA. Protein Ligands to HuR Modulate Its Interaction with Target mRNAs In Vivo. J Cell Biol 2000; 151:1-13.

129. Kwak H, Jeong KC, Chae MJ et al. Flavonoids inhibit the AU-rich element binding of HuC. BMB Rep 2009; 42:41-46.

130. Chae MJ, Sung HY, Kim EH et al. Chemical inhibitors destabilize HuR binding to the AU-rich element of TNF-alpha mRNA. Exp Mol Med 2009; 41:824-831.

131. Benoit RB, Meisner NC, Kallen J et al. The x-ray crystal structure on the first RNA recognition motif and site-corrected mutagenesis suggest a possible HuR redox sensing mechanism. J Mol Biol 2010; in press.

CHAPTER 11

TURNOVER OF MATURE miRNAs AND siRNAs IN PLANTS AND ALGAE

Heriberto Cerutti* and Fadia Ibrahim

Abstract: microRNAs (miRNAs) and small interfering RNAs (siRNAs) play important roles in gene regulation and defense responses against transposons and viruses in eukaryotes. These small RNAs generally trigger the silencing of cognate sequences through a variety of mechanisms, including RNA degradation, translational inhibition and transcriptional repression. In the past few years, the synthesis and the mode of action of miRNAs and siRNAs have attracted great attention. However, relatively little is known about mechanisms of quality control during small RNA biogenesis as well as those that regulate mature small RNA stability. Recent studies in *Arabidopsis thaliana* and *Caenorhabditis elegans* have implicated 3'-to-5' (SDNs) and 5'-to-3' (XRN-2) exoribonucleases in mature miRNA turnover and the modulation of small RNA levels and activity. In the green alga *Chlamydomonas reinhardtii*, a nucleotidyltransferase (MUT68) and an exosome subunit (RRP6) are involved in the 3' untemplated uridylation and the degradation of miRNAs and siRNAs. The latter enzymes appear to function as a quality control mechanism to eliminate putative dysfunctional or damaged small RNA molecules. Several post-transcriptional modifications of miRNAs and siRNAs such as 3' terminal methylation and untemplated nucleotide additions have also been reported to affect small RNA stability. These collective findings are beginning to uncover a new layer of regulatory control in the pathways involving small RNAs. We anticipate that understanding the mechanisms of mature miRNA and siRNA turnover will have direct implications for fundamental biology as well as for applications of RNA interference technology.

*Corresponding Author: Heriberto Cerutti—School of Biological Sciences and Center for Plant Science Innovation; University of Nebraska; P.O. Box 880666; Lincoln, Nebraska 68588, USA. Email: hcerutti1@unl.edu

Regulation of microRNAs, edited by Helge Großhans.
©2010 Landes Bioscience and Springer Science+Business Media.

INTRODUCTION

RNA-mediated silencing is an evolutionarily conserved mechanism(s) by which small RNAs (sRNAs) induce the inactivation of cognate sequences.[1-7] However, recent results indicate that these noncoding RNAs may also participate in transcriptional or translational activation.[2,8] The regulation of gene expression by sRNAs, ~20-30 nucleotides in length, plays an essential role in developmental pathways, metabolic processes and defense responses against viruses and transposons in many eukaryotes.[1-8] In plants and some algae, at least two major classes of small RNAs have been identified based on the molecules that trigger their production: microRNAs (miRNAs) and small interfering RNAs (siRNAs).[3-7,9-12] miRNAs originate from single-stranded noncoding RNA transcripts or introns that fold into imperfect stem-loop structures and often modulate the expression of genes with roles in development, physiological processes or stress responses.[4-7] siRNAs are produced from long, near-perfect complementarity double-stranded RNAs (dsRNAs) of diverse origins, including transcripts of long inverted repeats, products of convergent transcription or RNA-dependent RNA polymerase activity, viral and transposon RNAs, or dsRNAs experimentally introduced into cells.[4-7] In higher plants the siRNA population includes natural antisense transcript siRNAs (nat-siRNAs), trans-acting siRNAs (ta-siRNAs), heterochromatic siRNAs (hc-siRNAs), several other endogenous siRNAs (endo-siRNAs) as well as those derived from invading viral or transgene transcripts.[5-7,13] These siRNAs play various roles in post-transcriptional regulation of gene expression, suppression of viruses and transposable elements and/or DNA methylation and heterochromatin formation.[3-7,13] However, there is a growing realization that, despite their differences, distinct small RNA pathways often interact, competing for and sharing substrates, effector proteins and cross-regulating each other.

Hairpin and long dsRNAs are processed into small RNAs by an RNase III-like endonuclease named Dicer.[1,2,5,6] The short RNA duplexes produced by Dicer are incorporated into multisubunit effector complexes, such as the RNA-induced silencing complex (RISC).[1,2,5,6] Argonaute proteins, which include two main subfamilies of polypeptides named after *Arabidopsis thaliana* ARGONAUTE1 (AGO1) and *Drosophila melanogaster* PIWI, are core components of the RISC and some function as sRNA-guided endonucleases.[1-6,14,15] Recent evidence suggests that an siRNA duplex is first loaded into RISC and then AGO cleaves one of the siRNA strands (the passenger strand) triggering its dissociation from the complex.[1,2] Similarly, miRNA duplexes are loaded onto AGO and rapidly unwound by a poorly characterized mechanism.[2,16] Activated RISC then uses the remaining single-stranded small RNA as a guide to identify homologous RNAs, ultimately triggering transcript degradation and/or translational repression.[1-6] sRNAs associated with certain AGOs can also direct cytosine DNA methylation and/or chromatin modifications[4-7,13] and RISC complexes often contain auxiliary proteins that extend or modify their function(s).[1,2,8]

The biogenesis and the mode of action of sRNAs have attracted great attention,[1-8,17] but relatively little is known about mechanisms of mature miRNA/siRNA turnover and their role(s) in small RNA function. The accumulation of other cellular RNAs is dependent on the rates of transcription, processing and, also, decay. For instance, messenger RNA degradation is now known to contribute significantly to the post-transcriptional regulation of gene expression and as a quality control mechanism to prevent the expression of inappropriate RNAs.[18,19] By analogy, active small RNA turnover may conceivably modulate the levels of mature miRNAs/siRNAs and/or eliminate defective sRNA molecules. Here we examine the, as yet, relatively scant evidence on the mechanisms of small RNA degradation and their biological roles, with a specific focus on plants and algae. Along the way, we briefly review the biogenesis of miRNAs/siRNAs and seek to delineate the current knowns and the many unknowns in the field of small RNA turnover.

SMALL RNA PROCESSING

Most characterized eukaryotic miRNA genes correspond to RNA polymerase II transcription units, either in intergenic regions or embedded in introns of protein-coding genes, that produce a primary miRNA transcript (pri-miRNA).[1,2,5-7] This pri-miRNA typically forms an imperfect fold-back structure, which is processed into a short stem-loop precursor miRNA (pre-miRNA). In metazoans, this step is catalyzed by the nuclear microprocessor complex that includes as core components an RNase III enzyme (Drosha) and a double-stranded RNA-binding protein.[1,2,5,6,17] Pre-miRNAs are then exported to the cytoplasm by the karyopherin Exportin 5[20] and further processed in the cytosol by Dicer to generate mature miRNAs.[1,2,5,6,17] Dicer cleavage produces a short duplex containing two strands, named miRNA (equivalent to the guide strand) and miRNA* (the complementary, passenger strand).[1,2,5,6,17] In plants, which lack Drosha-like enzymes, both pri-miRNA-to-pre-miRNA conversion and pre-miRNA-to-duplex miRNA/miRNA* processing are carried out by Dicer-like proteins (Fig. 1).[1,5,6,13] In *Arabidopsis* these steps are largely dependent on the activity of the nuclear DICER LIKE 1 (DCL1).[5,6,21-23] However, higher plants and the green alga *Chlamydomonas reinhardtii* also have additional DCL proteins that are mostly responsible for the processing of a multitude of siRNAs from long dsRNAs, although they may also be involved in the making of some miRNAs.[4-7,11]

In metazoans, the biogenesis of certain miRNAs is regulated at the level of microprocessor-and/or Dicer-mediated processing, as demonstrated by the identification of RNA-binding proteins such as Lin-28, hnRNP A1 and KSRP that can either prevent or promote the conversion of specific pri-/pre-miRNAs to mature miRNAs (see chapters by Lehrbach and Miska, Michlewski et al, and Trabucchi et al, respectively).[17,24-26] Recently, the estrogen receptor α has also been implicated in inhibiting the processing of a subset of miRNAs that depend on the microprocessor-associated DEAD box helicases p68 and p72 for their biogenesis (see chapter by Fujiyama-Nakamura et al).[27] In addition, *Caenorhabditis elegans* and mammalian Lin-28, besides its role in pri-miRNA processing, can also bind the precursor of the let-7 miRNA in the cytoplasm and stimulate its 3′ end uridylation by a poly(U) polymerase, leading to precursor RNA degradation and downregulation of the mature let-7 miRNA levels (see chapter by Lehrbach and Miska).[28-30] In contrast to this wealth of information, to our knowledge, there is as yet no experimental evidence supporting miRNA-specific regulation at the processing steps in plants or algae. However, discrepancies between pri-/pre-miRNA and mature miRNA levels in northern blot analyses of certain miRNAs suggest that post-transcriptional mechanisms affecting miRNA accumulation are also likely to exist in plants.[5,31]

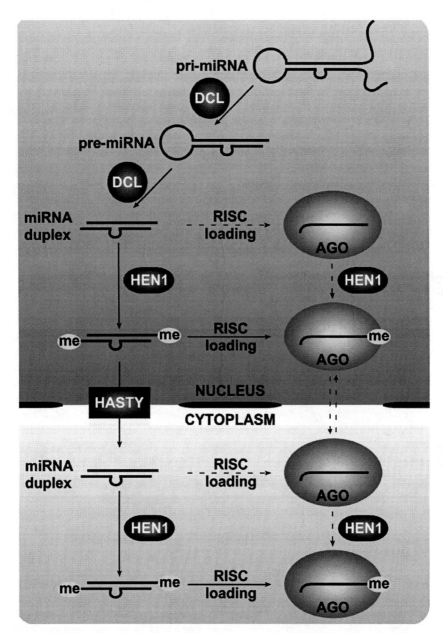

Figure 1. General model of miRNA biogenesis and RISC loading in plants and some algae. Primary miRNA transcripts (pri-miRNAs), mostly generated by RNA polymerase II, are processed into hairpin precursor miRNAs (pre-miRNAs) by Dicer-like enzymes (DCL). These pre-miRNAs are further processed by DCL into short miRNA/miRNA* duplexes (miRNA duplexes). Mature miRNA duplexes are then methylated at the 3' end of each strand by HEN1. Some miRNA/miRNA* or methylated miRNA/ miRNA* duplexes are likely exported to the cytoplasm by HASTY, the plant homolog of Exportin 5. Guide miRNA strands are eventually loaded, either in the nucleus or the cytosol, into effector complexes containing Argonautes (AGOs). Commonly accepted pathways in higher plants are indicated with solid lines whereas potential alternative pathways, currently lacking direct experimental evidence, are indicated with dashed lines (see text for details).

SMALL RNA MODIFICATION BY 2'-O-METHYLATION

In plants, mature miRNAs and siRNAs are methylated at their 3' ends, a modification dependent on the RNA methyltransferase HUA ENHANCER 1 (HEN1).[5,6,13,32] This is also likely to occur in the alga *C. reinhardtii*, as suggested by the resistance of its small RNAs to periodate oxidation/β elimination reactions.[9,33] In vitro studies with recombinant HEN1 strongly suggest that the *Arabidopsis* protein prefers as substrates small RNA duplexes with 2-nt overhangs at their 3' ends, typical features of Dicer products.[5,6,13,34,35] Thus, after DCL proteins catalyze the release of miRNA/miRNA* or siRNA duplexes from their precursors, it has been proposed that HEN1 methylates each strand of the duplex on the 2'-OH of their 3'-terminal ribose molecules.[6,13,34] Interestingly, a HEN1-YFP fusion protein has been detected in both the nucleus and the cytosol in transgenic *Arabidopsis* lines[21] and several viral RNA silencing suppressors, that appear to function in the cytoplasm, partly inhibit miRNA methylation.[36] Thus, it seems likely that HEN1-catalyzed reactions can occur in the nucleus as well as in the cytosol of plant cells (Fig. 1), although this has not been formally demonstrated.

In metazoans, PIWI-interacting RNAs (piRNAs), a class of small RNAs specifically bound by PIWI proteins and absent in plants, as well as several siRNAs also have a 2'-O-methyl group on their 3' termini.[37-40] In flies, miRNA*s associated with AGO2 have also been found to be 3'-modified.[41-43] In contrast, animal miRNAs do not appear to be methylated.[1,37-40] Moreover, the animal homologs of HEN1 lack a dsRNA-binding domain and appear to act on single-stranded, mature small RNAs already associated with AGO or PIWI proteins.[38-40] Indeed, the substrate specificity of HEN1 homologs in metazoans may reflect the fact that these proteins only interact with certain Argonaute polypeptides. Whether plant HEN1 could also methylate some single-stranded small RNAs already bound to AGOs is presently unknown (Fig. 1, dashed lines pathway). In both animals and plants, the methylation of small RNAs seems to protect them against untemplated nucleotide additions, such as uridylation, and/or exonucleolytic shortening.[6,13,38,39,44] Likewise, in the ciliated protozoan *Tetrahymena thermophila*, ~28-29 nt long sRNAs, which are expressed during sexual reproduction and required for DNA elimination, are selectively stabilized by 3'-terminal 2'-O-methylation.[45]

SMALL RNA LOADING AND ACTIVATION OF THE RNA-INDUCED SILENCING COMPLEX

In metazoans, siRNAs seem to be loaded onto RISC as duplexes and then AGO cleaves the passenger strands triggering their dissociation from the complex and the concomitant maturation of RISC.[1,2,17,46] Ribonucleases, such as C3PO (whose subunits Translin and Translin-associated factor X have homologs in plants), promote RISC activation by removing the passenger strand cleavage products.[46] Likewise, miRNA/ miRNA* duplexes, which often contain mismatches or bulges, are loaded onto AGO and the two strands are separated by a poorly defined "slicer-independent" mechanism.[2,16] However, recent evidence suggests that AGO proteins themselves can function as RNA chaperones capable of unwinding small RNA duplexes.[16,47] The dissociated miRNA* strands appear to be rapidly degraded, but the enzyme(s) involved in this process is presently unknown (Fig. 2). In either case, no single-stranded guide siRNA or miRNA appears to be produced prior to these RISC maturation steps.[2,38,48]

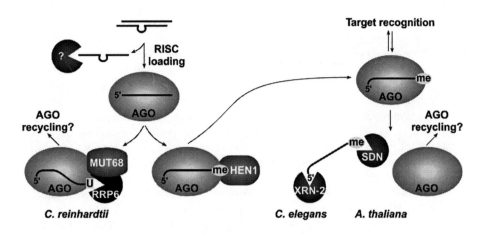

Figure 2. Proposed model for the turnover of mature miRNAs based on combined evidence from plants, algae and metazoans. An miRNA/miRNA* duplex is loaded into AGO and the two strands are separated by a poorly characterized mechanism. The unwound miRNA* strand is rapidly eliminated by an unknown enzyme(s). In *C. reinhardtii*, MUT68 (a terminal nucleotidyltransferase) and RRP6 (a 3'-to-5' exoribonuclease) may be part of a quality control mechanism to eliminate dysfunctional or damaged miRNAs that are loaded into Argonautes, in kinetic competition with the methyltransferase HEN1 (see text for details). Both *C. elegans* XRN-2 (a 5'-to-3' exoribonuclease) and *A. thaliana* SDN enzymes (3'-to-5' exoribonucleases) may contribute to the decay of mature small RNAs dissociated from Argonautes (see text for details). For simplicity, the AGO-bound miRNA strand is shown methylated, as it occurs in plants and some algae, but *C. elegans* miRNAs would not undergo this modification.

In several metazoans, the relative thermodynamic stability of the 5' ends of the strands in a small duplex RNA, in some cases sensed by dsRNA-binding proteins partnering with Dicer in a RISC-loading complex, determines which strand is chosen as the guide siRNA or miRNA.[2,16,41,49,50,51] In addition, in *Drosophila*, sorting of small RNA duplexes into specific AGO paralogs appears to be governed by the structure of the duplex, whether it is nearly perfectly double-stranded or contains central bulges and mismatches.[16,41,43] As reported in plants, the identity of the first nucleotide of a small RNA may also play a role in this sorting process.[41,43] However, the extent to which these factors weigh in the fate of specific small RNAs, in particular in different metazoans, is not clear as yet.[52,53] Moreover, it has been commonly accepted that the passenger and miRNA* strands are simply byproducts of siRNA/miRNA biogenesis and RISC loading, destined to be degraded. Yet, recent evidence in flies suggests that certain miRNA/miRNA* duplexes could be bifunctional, with each strand being independently sorted into different AGO proteins and most miRNA*s detected in cells appear to represent those associated with Argonaute proteins rather than undegraded discarded strands.[41-43]

Much less is known about RISC assembly in plants and algae, and elucidating this process is complicated by the existence of many Argonaute paralogs in a given organism.[4-7,11,13,54] In *A. thaliana*, which contains ten AGOs, some heterochromatic and repetitive siRNAs are loaded into AGO4-containing complexes, likely in the nucleus.[5,13,55-57] In contrast, most miRNAs appear to become associated with AGO1.[5,6,54,56-58] At least part of the sorting into different *Arabidopsis* AGOs seems to be determined by the identity of the 5' nucleotide of the small RNAs.[56,57,59] For instance, AGO1 predominantly associates with small RNAs with a uridine at the 5' terminus, which most miRNAs possess, whereas

AGO4 prefers an adenine as the 5'-terminal nucleotide.[5,56,57] Additionally, the asymmetric thermodynamic stability of the miRNA/miRNA* duplex termini also appears to play a role in miRNA strand selection in plants, but these rules do not seem to apply to at least some siRNAs.[51]

The subcellular location of miRNA loading into AGOs remains elusive in plants. HASTY (HST), the plant homolog of Exportin 5, is thought to transport miRNA/miRNA* or methylated miRNA/miRNA* duplexes to the cytoplasm[5,6,60] for assembly into AGO complexes (Fig. 1). However, the role of HST is not as clear as that of Exportin 5 in animals since *Arabidopsis hasty* mutants show decreased accumulation of only a subset of miRNAs.[5,60] Moreover, in plants, miRNA abundance is higher than that of the corresponding miRNA* in both the cytoplasm and the nucleus, suggesting that mature, RISC-associated miRNAs are present in both compartments.[60] Interestingly, *Arabidopsis* HYPONASTIC LEAVES 1/DsRNA-BINDING PROTEIN 1 (HYL1/DRB1), a dsRNA-binding protein that cooperates with DCL1 in the processing of pri-/pre-miRNAs to mature miRNAs,[61] influences miRNA strand selection, presumably in a similar way as related polypeptides in some metazoan RISC loading complexes.[51] Since *Arabidopsis* HYL1/DRB1 is mainly localized in the nucleus[21,22] and a YFP-AGO1 fusion protein is present in both the cytosol and the nucleus,[21] at least a subset of AGO1 molecules may interact with HYL1/DRB1 and be loaded with miRNA/miRNA* or methylated miRNA/miRNA* duplexes in the nuclear compartment (Fig. 1). Recent findings in mammalian cells also indicate that Argonaute proteins and associated miRNAs can shuttle between the nucleus and the cytoplasm and that their transport depends on the import receptor Importin 8 and the karyopherin CRM1.[62,63]

Another unresolved issue in plant RISC assembly is the exact role of the methyltransferase HEN1, proposed to act on short dsRNA substrates after DCL-mediated processing but prior to RISC loading.[6,13,34] For instance, small RNA duplexes generated by DCL could be released, methylated by HEN1 and then rebound by metazoan-like RISC-loading complexes that associate with Argonautes.[51] Alternatively, HEN1 could be an integral component of plant RISC-loading complexes and participate actively in the transfer of small RNAs to AGOs.

MATURE SMALL RNA DEGRADATION BY RIBONUCLEASES

Relatively little is known about the stability of endogenous small RNAs and the enzymes involved in their turnover in most eukaryotes. A conserved nuclease from *C. elegans* and *Schizosaccharomyces pombe*, ERI-1 (of which there are also six putative homologs in *Arabidopsis*),[64] degrades siRNA duplexes with 2-nucleotide 3'-overhangs in vitro and reduces the efficiency of RNAi in vivo.[65,66] However, its role in small RNA turnover is not clear since, in nematodes, ERI-1 has recently been implicated in 5.8S rRNA processing and in the biogenesis of certain endo-siRNAs.[67,68] In contrast, in *C. elegans*, the 5'-to-3' exoribonuclease XRN-2 (related to the yeast Rat1 enzyme) is involved in the degradation of mature, single-stranded miRNAs (Fig. 2) and has been shown to modulate miRNA accumulation in vivo (see chapter by Grosshans and Chatterjee).[69]

In *Arabidopsis*, the existence of ribonucleases targeting siRNAs/miRNAs and the protective role of the 3'-terminal 2'-*O*-methyl group was recognized from analyses of small RNAs in mutants lacking sRNA methyltransferase activity.[6,13,32,44] In *hen1* mutants, miRNAs and siRNAs fail to accumulate or their levels are considerably reduced.[13,44] In addition, miRNA cloning and sequencing revealed the presence of 3' end-truncated miRNA molecules as well as others with untemplated 3'-terminal nucleotides, predominantly uridine residues.[13,44] These observations indicated that methylation protects small RNAs from uridylation and degradation and, by analogy to the mechanism of decay of longer transcripts such as human histone mRNAs,[70,71] led to the proposal that uridylation recruits and/or stimulates an exonuclease to degrade miRNAs.[6,13] Interestingly, a family of 3'-to-5' exoribonucleases (related to the yeast Rex exonucleases) encoded by the *SMALL RNA DEGRADING NUCLEASE* (*SDN*) genes was recently implicated in the turnover of single-stranded, mature sRNA in *Arabidopsis*[64] (Fig. 2). However, the enzymes involved in untemplated nucleotide additions to the 3' ends of mature sRNAs and in the proposed 3'-to-5' degradation of unmethylated and uridylated small RNAs remain unknown since SDN1 is inhibited by 3' terminal uridylation while it still acts, albeit with somewhat lower efficiency, on 2'-*O*-methylated sRNAs.[64]

Recent studies with in vitro systems (either cell extracts or recombinant proteins) demonstrated that single-stranded, guide small RNAs can be dissociated from Argonaute proteins.[47,69] In *C. elegans* extracts, this process is partly dependent on XRN-2 and appears to be inhibited by interaction of the miRNA-AGO complex with a target RNA.[69] If confirmed in vivo, this mechanism could provide a way to recycle AGO proteins associated with sRNAs that lack a target transcript, allowing them to rebind to other guide siRNAs/miRNAs. Nevertheless, current evidence is most consistent with both *C. elegans* XRN-2[69] and *Arabidopsis* SDN enzymes[64] participating in the decay of mature small RNAs dissociated from Argonautes (Fig. 2). Moreover, since homologs of these proteins are widely distributed among eukaryotes,[64,69,72] these pathways might be evolutionarily conserved; although partly redundant, multiple paralogs may complicate the detection of phenotypic defects in individual mutants or epi-mutants.[64,69] Alternatively, the prevalence of 5'-to-3' versus 3'-to-5' degradation of dissociated mature small RNAs may vary in different organisms since *C. elegans* homologs of *Arabidopsis* SDN1 do not appear, individually, to be required for miRNA turnover[69] and the *Arabidopsis* XRN-2 homologs XRN2 and XRN3 seem to degrade the loop sequence of miRNA precursors without affecting mature miRNA levels.[73]

A mutant in the green alga *Chlamydomonas reinhardtii* (Mut-68) also provided insight on the pathways of mature miRNA/siRNA degradation. Mut-68, which is deleted for a gene encoding a terminal nucleotidyltransferase named MUT68, was initially characterized as being deficient in the addition of untemplated nucleotides to the 5' RNA fragments produced by the RISC cleavage of target transcripts, a requirement for their efficient decay.[74] In addition, Mut-68 showed elevated levels of miRNAs and siRNAs and the MUT68 enzyme was found to play a role in the untemplated uridylation of the 3' termini of sRNAs in *Chlamydomonas*.[33] High throughput sequencing of small RNAs revealed that ~7.3% of the examined molecules had 3'-untemplated nucleotides in the wild type strain but this fraction was reduced to ~4.9% in Mut-68. Moreover, sRNAs displayed markedly lower uridylation, the predominant addition to the 3' ends of miRNAs/siRNAs, in the mutant and, consistent with the possibility that U-tailed RNAs may be degradation intermediates, their average size was smaller than that of the sRNAs in the entire population.[33]

The MUT68 activity stimulated in vitro the degradation of single-stranded small RNAs by RRP6,[33] a peripheral component of a 3'-to-5' multisubunit exoribonuclease, the exosome.[75,76] Moreover, like the defect in MUT68, RNAi-mediated depletion of RRP6 in *Chlamydomonas* resulted in the accumulation of miRNAs and siRNAs in vivo.[33] RRP6, which is related to bacterial RNase D, is widely distributed in eukaryotes and acts as a distributive 3'-to-5' hydrolytic exonuclease that prefers unstructured substrates.[75,77] As proposed before, it seems likely that, in *Chlamydomonas*, uridylation by MUT68 creates a short unstructured 3' end that facilitates small RNA degradation by the RRP6 enzyme (Fig. 2). Several cycles of uridylation and truncation may be required for complete sRNA decay by this nonprocessive exoribonuclease. Interestingly, MUT68 appears to collaborate with RRP6 in the turnover of miRNAs/siRNAs[33] and with the core exosome in the degradation of longer RNAs generated by RISC cleavage.[74] Additionally, MUT68 seems to carry out preferentially uridylation of small RNAs[33] and adenylation of RISC-cleaved transcripts.[74] The basis for this differential specificity is presently unclear but nucleotidyltransferases with context-dependent nucleotide preferences have been previously described.[70,78,79] Furthermore, in respect to sRNA degradation, 3'-terminal adenylation, unlike uridylation, has recently been proposed to lead to stabilization of miRNAs. The poly(A) polymerase GLD-2 adds a single adenine residue to the 3' end of mammalian miR-122 and this modification appears to stabilize selectively this particular miRNA in liver cells.[80] Untemplated adenylation of

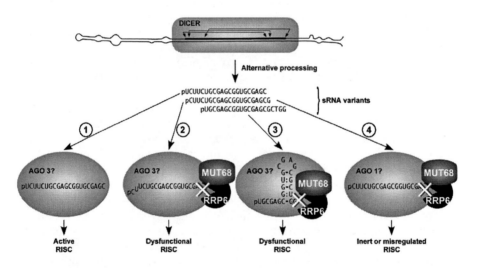

Figure 3. Proposed model for the role of MUT68 and RRP6 in the quality control of mature small RNAs in *Chlamydomonas reinhardtii*. Slight errors during Dicer-mediated processing and/or cleavage by alternative Dicer paralogs result in 5' nucleotide variants of at least some miRNAs and endo-siRNAs.[33] In the wild type strain one sRNA isoform usually predominates and is presumably associated with an active RISC (1). However, in the Mut-68 mutant alternative isoforms can also become quite noticeable,[33] suggesting that their decay depends on the MUT68 nucleotidyltransferase activity. Certain 5' nucleotide variants may be subfunctional or dysfunctional when associated with a particular Argonaute protein, for instance because of an unsuitable 5' terminal nucleotide (2) and/or a tendency to form intramolecular secondary structures (3) that will hinder the recognition of target RNAs. In the absence of the MUT68/RRP6 machinery, these small RNA variants can accumulate, conceivably sequestering AGO proteins into inactive RISC complexes. Additionally, some 5' nucleotide isoforms of miRNAs/siRNAs may be loaded into different AGO paralogs (4). In strains depleted for MUT68 and/or RRP6, these small RNAs may compete out those commonly associated with these Argonaute proteins, rendering the complexes functionally inert or leading to altered regulatory outcomes. Since small RNA processing isoforms have different seed sequences, if assembled into functional complexes, they could potentially affect the expression of distinct repertoires of target RNAs. For simplicity, MUT68 and RRP6 are shown degrading RISC-associated small RNAs but this has not been directly demonstrated as yet.

miRNAs has also been observed in plants and algae[33,44,81] and it also seems to protect small RNAs against degradation in an in vitro assay with *Populus trichocarpa* (black cottonwood) cell extracts.[81]

Both MUT68 and RRP6 are only active in vitro on small RNAs lacking a 3'-terminal 2'-*O*-methyl group.[33] Thus, homologs of MUT68 and RRP6 may conceivably be responsible for the observed uridylation and decay of small RNAs lacking 3' methylation in the *Arabidopsis hen1* mutants. However, defining the role(s) of MUT68 and RRP6 in *Arabidopsis* may be complicated by the fact that both proteins are encoded by small multigene families.[74,82] More importantly, how these enzymes function in a wild-type background, where most miRNAs and siRNAs are methylated, is less obvious. As discussed in the next section, we have proposed[33] that, at least in *Chlamydomonas*, MUT68 and RRP6 may be part of a quality control mechanism to eliminate dysfunctional or damaged small RNAs associated with Argonautes (Fig. 3).

QUALITY CONTROL OF MATURE SMALL RNAs

The *Chlamydomonas* Mut-68 mutant was originally identified as being deficient in RNAi[74] and the RRP6-depleted strains also shows diminished RNAi activity. However, since Mut-68 contains enhanced levels of mature, single-stranded miRNAs and siRNAs, which correlate with higher amounts of an endogenous AGO protein, RISC assembly appears to occur normally.[33] As already mentioned, no single-stranded siRNA or miRNA appears to be produced prior to RISC maturation[2,38,48] and, thus, the accumulated mature sRNAs detected in *Chlamydomonas* Mut-68 likely correspond to those associated with Argonautes. Yet, the function of a significant fraction of these RISC complexes may be compromised if the associated guide sRNAs are dysfunctional, inert and/or damaged, resulting in the sequestration of AGO proteins into inactive complexes (Fig. 3). This interpretation for the diminished RNAi activity in Mut-68 (and in the RRP6-depleted strains) is consistent with a role for MUT68/RRP6 as a quality control mechanism for the removal of functionally defective sRNAs in *Chlamydomonas* (Fig. 3). Moreover, this process may be operative in other eukaryotes since a recent RNAi screen to identify genes involved in miRNA/siRNA pathways in *D. melanogaster* revealed that depletion of an RRP6 homolog resulted in an RNAi defect.[83] In addition, the *C. elegans* nucleotidyltransferase CDE-1 is required for the uridylation of siRNAs bound to a specific Argonaute protein (CSR-1) and in the absence of CDE-1 these siRNAs accumulate to inappropriate levels, accompanied by defects in an RNAi pathway involved in chromosome segregation.[84]

Recent evidence suggests that RISC-bound small RNAs can be subfunctional. For instance, changing the 5' uracil residue of the let-7a miRNA did not affect the formation of a complex with human AGO2 but reduced significantly the association of this complex with a target mRNA.[85] In *Arabidopsis*, a uridine-to-adenosine change at the 5' end of engineered miRNAs resulted in an AGO1-to-AGO2 switch in sRNA loading and abolished their silencing activity.[56] In plants and some algae, slight errors during DCL processing and/or cleavage by alternative DCL paralogs may result in 5' nucleotide variants of miRNAs/siRNAs that could be assembled into the wrong AGO isoform and have drastically altered regulatory outcomes,[5,86] including rendering the miRNA/siRNA functionally inert and sequestering Argonaute proteins into ineffective complexes. Inaccuracies by RISC-loading complexes may also lead to the association of small RNAs with an incorrect AGO paralog. Thus, a quality control mechanism(s) may be required to eliminate AGO-bound dysfunctional or subfunctional small RNAs and MUT68 and RRP6 may participate in such a pathway (Fig. 3). We have not demonstrated directly that MUT68 and RRP6 act on Argonaute-associated small RNAs but, in *C. elegans*, uridylated siRNAs are immunoprecipitated with the CSR-1 AGO.[84] In addition, both *Chlamydomonas* Mut-68 and CDE-1-defective *C. elegans* are deficient in RNA interference pathways, suggesting that, in these mutants, the accumulated small RNAs hinder RISC activity.[33,84] In contrast, this phenotype has not been reported upon depletion of *C. elegans* XRN-2 or *Arabidopsis* SDNs, implicated in the turnover of small RNAs dissociated from Argonautes.[64,69] One expectation is that the populations of mature small RNAs accumulated in these sets of mutants would be different, including sRNAs with processing defects (and/or associated with incorrect AGOs) in the first case and predominantly correctly processed, functional miRNAs/siRNAs in the second.

Chlamydomonas MUT68/RRP6 may function in competition with HEN1 in a putative assessment of small RNA functionality. The *D. melanogaster* HEN1 homolog appears to methylate single-stranded piRNAs and siRNAs already associated with certain AGO/PIWI proteins.[1,38,39] Thus, sRNAs lacking 2'-*O*-methyl groups are loaded into RISC in animals and, conceivably, this may also happen for at least a fraction of the small RNAs in *Chlamydomonas*. We proposed that, in these cases, the MUT68/RRP6 machinery may operate as a quality control mechanism in kinetic competition with HEN1 (Fig. 2).[33] Functional guide sRNAs (with respect to their interactions with a particular AGO isoform) may be protected by HEN1-mediated 3' end methylation whereas subfunctional or dysfunctional sRNAs may be preferentially degraded by MUT68/RRP6 (Fig. 2). However, it is not clear whether a similar mechanism could also act in higher plants where HEN1 has been suggested to methylate small RNA duplexes prior to their loading into RISC.[6,13,34] Additionally, dysfunctional or subfunctional small RNAs may be conceivably dissociated more easily from an Argonaute protein and XRN-2 and/or SDN homologs could also contribute to the degradation of some of these molecules.

In degradative RNAi, RISC functions as a multiple turnover enzyme[2,87] and a quality control mechanism(s) may also be necessary to assess the integrity of guide siRNAs after each round of target RNA cleavage. In mature RISC, the 3' end of the guide siRNA is bound by the AGO PAZ domain but, when the siRNA forms an extensive duplex with a target RNA, its 3' terminus is released from the PAZ pocket.[14,15,88] After RISC-mediated endonucleolytic cleavage, the target RNA products are released and degraded by exoribonucleases.[74,89,90] At this step, the 3' end of the guide siRNA may become accessible to the MUT68/RRP6 machinery prior to rebinding to the PAZ domain. We speculated that MUT68/RRP6 may also operate here, as a quality control mechanism to degrade damaged sRNAs lacking 2'-*O*-methyl groups.[33] However, understanding the molecular details of the proposed quality control mechanism(s) will require addressing the nature of the putative dysfunctional or damaged small RNAs.

CONCLUSION

Small RNAs, both miRNAs and siRNAs, play important roles in the regulation of gene expression in eukaryotes. Yet, the complexity of small RNA biogenesis and function is just beginning to be understood. Recent studies have established that post-transcriptional sRNA modifications (such as 3' terminal methylation and untemplated nucleotide additions) and several exoribonucleases can affect the stability of mature, single-stranded miRNAs and siRNAs.[33,64,69,91] Moreover, these factors can have profound effects on the homeostasis and the function of small RNAs in plants, algae and metazoans.[33,64,69,91] However, despite these advances, we still know relatively little about the molecular mechanisms of mature small RNA turnover and whether the discovered pathways are common to most eukaryotes. Additionally, many enzymes implicated in the modification and the degradation of miRNAs and/or siRNAs appear to be encoded by small multigene families, in plants, algae and animals, and potential redundancy of function may complicate uncovering their significance in sRNA metabolism and developmental and physiological responses.

The biological role(s) of small RNA degradation also needs further exploration. Some pathways may operate as quality control mechanisms to eliminate AGO-associated dysfunctional or subfunctional small RNAs, resulting from errors in Argonaute loading and/or mistakes in the processing of miRNAs/siRNAs.[5,33] Critical questions in this context are the nature of the postulated dysfunctional or subfunctional small RNAs and the way they are recognized by the degradation machinery. Other turnover mechanisms may modulate the overall levels of AGO-bound miRNAs.[64,69] An intriguing possibility raised by work in *C. elegans* is that the accumulation of small RNAs may be linked to the availability of target RNAs,[69] provided that target binding maintains miRNAs in an Argonaute-associated state protected from exonuclease-mediated degradation. These pathways would potentially facilitate the recycling of AGO proteins in dysfunctional or inert complexes for rebinding to other small RNAs. Whether the levels of specific mature miRNAs could also be regulated by selective turnover is not clear as yet. In this case, factors that recognize certain miRNA sequences would presumably be needed to recruit ribonucleases to particular substrates. Interestingly, a 3' terminal hexanucleotide sequence in human miR-29b promotes its nuclear localization, suggesting that specific small RNA sequence motifs can direct distinct outputs, but the factors involved in this selective localization are not known.[92]

Untemplated nucleotide additions appear to influence the stability of mature miRNAs and siRNAs but the significance of small RNA modifications is not entirely obvious. For instance, 3'-terminal uridylation may create unstructured sRNA ends, facilitating their degradation by nonprocessive exoribonucleases.[13,28,33,71] However, 3' end uridylation of mature miR-26 in mammalian cells appears to impart functional differences that attenuate miRNA-targeted repression without noticeable changes in miRNA steady-state levels.[93] Likewise, in *Arabidopsis* certain 3'-uridylated miRNAs are almost as abundant as the unmodified canonical forms in particular tissues, suggesting a specialized role for these modified small RNAs.[86] The 3'-terminal adenylation of some miRNAs appears to stabilize them[80,81] but this same modification promotes the degradation of the 5' RNA products of RISC cleavage and a number of misprocessed and unstable RNAs.[74-76,94-96] Thus, the consequences of untemplated nucleotide additions to small RNAs may be context-dependent. Conceivably, the effect of 3'-untemplated nucleotide additions may depend on how they alter the length and/or the 3'-terminal structure of a given miRNA or siRNA and, as a result, the sRNA interactions with AGO and susceptibility to ribonuclease activities.

Similarly, 3' end methylation of small RNAs seems to protect them directly from nucleotidyltransferases and exoribonucleases,[13,38,39,44,45] but additional functions for this modification have not been explored. For instance, 3'-terminal methylation may affect the association of an sRNA with the PAZ domain of AGO, potentially influencing the kinetics of double-strand zippering with a target RNA. Indeed, a 2'-O-methyl group on the 3'-terminal nucleotide appears to decrease the siRNA-binding affinity by the PAZ domain of human AGO1.[97] In higher plants, 2'-O-methylation may also promote or decrease the ability of RNA-dependent RNA polymerases to use small RNAs as primers.[13] Another outstanding question is why all siRNAs and miRNAs are methylated in plants (and likely in some algae) whereas miRNAs do not seem to undergo this modification in metazoans. Interestingly, it has been noted that all 2'-O-modified small RNAs identified thus far are associated with RISC complexes that have the capability to cleave efficiently their RNA targets.[38] This might reflect a requirement of 3' methylation of sRNAs (through its potential effect on AGO binding) for optimal duplex formation as an intermediate for target RNA cleavage and/or a role of this modification in preventing the unintended degradation of small RNAs by the exoribonucleases that participate in the decay of RISC-cleaved RNA products.

Finally, the subcellular localization of the pathways that affect small RNA stability remains to be evaluated in most eukaryotes. In *Chlamydomonas*, MUT68 appears to be located predominantly in the cytosol,[33] but it is becoming increasingly clear that distinct RISC complexes function in both the nucleus and the cytoplasm.[1-8] Thus, certain pathways for small RNA degradation, for instance those associated with quality control, may be required to operate in both compartments. Conversely, selective subcellular localization of turnover processes might provide another layer of regulation for the degradation of specific miRNAs. We anticipate that deepening our knowledge about the mechanisms that regulate mature small RNA turnover will be relevant not only to the comprehensive understanding of how miRNA and siRNAs execute their function but also to the successful use of RNAi for practical applications.

ACKNOWLEDGMENTS

This work was supported by a grant from the National Institutes of Health to H.C. We also acknowledge the support of the Nebraska EPSCoR program.

REFERENCES

1. Ghildiyal M, Zamore PD. Small silencing RNAs: an expanding universe. Nat Rev Genet 2009; 10:94-108.
2. Carthew RW, Sontheimer EJ. Silence from within: endogenous siRNAs and miRNAs. Cell 2009; 136:642-55.
3. Baulcombe D. RNA silencing in plants. Nature 2004; 431:356-63.
4. Cerutti H, Casas-Mollano JA. On the origin and functions of RNA-mediated silencing: from protists to man. Curr Genet 2006; 50:81-99.
5. Voinnet O. Origin, biogenesis and activity of plant microRNAs. Cell 2009; 136:669-87.
6. Chen X. Small RNAs and their roles in plant development. Annu Rev Cell Dev Biol 2009; 35:21-44.
7. Chapman EJ, Carrington JC. Specialization and evolution of endogenous small RNA pathways. Nat Rev Genet 2007; 8:884-96.
8. Steitz JA, Vasudevan S. miRNPs: versatile regulators of gene expression in vertebrate cells. Biochem Soc Trans 2009; 37:931-35.
9. Molnár A, Schwach F, Studholme DJ et al. miRNAs control gene expression in the single-cell alga Chlamydomonas reinhardtii. Nature 2007; 447:1126-29.
10. Zhao T, Li G, Mi S et al. A complex system of small RNAs in the unicellular green alga Chlamydomonas reinhardtii. Genes Dev 2007; 21:1190-203.
11. Casas-Mollano JA, Rohr J, Kim EJ et al. Diversification of the core RNA interference machinery in Chlamydomonas reinhardtii and the role of DCL1 in transposon silencing. Genetics 2008; 179:69-81.

12. De Riso V, Raniello R, Maumus F et al. Gene silencing in the marine diatom Phaeodactylum tricornutum. Nucleic Acids Res 2009; 37:e96.
13. Ramachandran V, Chen X. Small RNA metabolism in Arabidopsis. Trends Plant Sci 2008; 13:368-74.
14. Yuan YR, Pei Y, Ma JB et al. Crystal structure of A. aeolicus argonaute, a site-specific DNA-guided endoribonuclease, provides insights into RISC-mediated mRNA cleavage. Mol Cell 2005; 19:405-19.
15. Wang Y, Juranek S, Li H et al. Nucleation, propagation and cleavage of target RNAs in Ago silencing complexes. Nature 2009; 461:754-61.
16. Kawamata T, Seitz H, Tomari Y. Structural determinants of miRNAs for RISC loading and slicer-independent unwinding. Nat Struct Mol Biol 2009; 16:953-60.
17. Winter J, Jung S, Keller S et al. Many roads to maturity: microRNA biogenesis pathways and their regulation. Nat Cell Biol 2009; 11:228-34.
18. Wilusz CJ, Wilusz J. Bringing the role of mRNA decay in the control of gene expression into focus. Trends Genet 2004; 20:491-97.
19. Isken O, Maquat LE. Quality control of eukaryotic mRNA: saferguarding cells from abnormal mRNA function. Genes Dev 2007; 21:1833-56.
20. Okada C, Yamashita E, Lee SJ et al. A high-resolution structure of the pre-microRNA nuclear export machinery. Science 2009; 326:1275-1279.
21. Fang Y, Spector DL. Identification of nuclear dicing bodies containing proteins for miRNA biogenesis in living Arabidopsis plants. Curr Biol 2007; 17:818-23.
22. Song L, Han MH, Lesicka J et al. Arabidopsis primary microRNA processing proteins HYL1 and DCL1 define a molecular body distinct from the Cajal body. Proc Natl Acad Sci USA 2007; 104:5437-42.
23. Fujioka Y, Utsumi M, Ohba Y et al. Location of a possible miRNA processing site in SmD3/SmB nuclear bodies in Arabidopsis. Plant Cell Physiol 2007; 48:1243-53.
24. Michlewski G, Guil S, Semple CA et al. Post-transcriptional regulation of miRNAs harboring conserved terminal loops. Mol Cell 2008; 32:383-93.
25. Viswanathan SR, Daley GQ, Gregory RI. Selective blockade of microRNA processing by Lin28. Science 2009; 320:97-100.
26. Trabucchi M, Briata P, Garcia-Mayoral M et al. The RNA-binding protein KSRP promotes the biogenesis of a subset of microRNAs. Nature 2009; 459:1010-4.
27. Yamagata K, Fujiyama S, Ito S et al. Maturation of microRNA is hormonally regulated by a nuclear receptor. Mol Cell 2009; 36:340-7.
28. Heo I, Joo C, Kim Y-K et al. TUT4 in concert with Lin28 suppresses microRNA biogenesis through pre-microRNA uridylation. Cell 2009; 138:696-708.
29. Hagan JP, Piskounova E, Gregory RI. Lin28 recruits the TUTase Zcchc11 to inhibit let-7 maturation in mouse embryonic stem cells. Nat Struct Mol Biol 2009; 16:1021-25.
30. Lehrbach NJ, Armisen J, Lightfoot HL et al. LIN-28 and the poly(U) polymerase PUP-2 regulate let-7 microRNA processing in Caenorhabditis elegans. Nat Struct Mol Biol 2009; 16:1016-20.
31. Nogueira F, Chitwood D, Madi S et al. Regulation of small RNA accumulation in the maize shoot apex. PLoS Genet 2009; 5:e1000320.
32. Yu B, Yang Z, Li J et al. Methylation as a crucial step in plant microRNA biogenesis. Science 2005; 307:932-35.
33. Ibrahim F, Rymarquis LA, Kim E-J et al. Uridylation of mature miRNAs and siRNAs by the MUT68 nucleotidyltransferase promotes their degradation in Chlamydomonas. Proc Natl Acad Sci USA 2010; 107:3906-11.
34. Yang Z, Ebright YW, Yu B et al. HEN1 recognizes 21-24 nt small RNA duplexes and deposits a methyl group onto the 2' OH of the 3' terminal nucleotide. Nucleic Acids Res 2006; 34:667-75.
35. Huang Y, Ji L, Huang Q et al. Structural insights into mechanisms of the small RNA methyltransferase HEN1. Nature 2009; 461:823-27.
36. Yu B, Chapman EJ, Yang Z et al. Transgenically expressed viral RNA silencing suppressors interfere with microRNA methylation in Arabidopsis. FEBS Lett 2006; 580:3117-20.
37. Farazi TA, Juranek SA, Tuschl T. The growing catalog of small RNAs and their association with distinct Argonaute/Piwi family members. Development 2008; 135:1201-14.
38. Horwich MD, Li C, Matranga C et al. The Drosophila RNA methyltransferase, DmHen1, modifies germline piRNAs and single-stranded siRNAs in RISC. Curr Biol 2007; 17:1265-72.
39. Saito K, Sakaguchi Y, Suzuki T et al. Pimet, the Drosophila homolog of HEN1, mediates 2'-O-methylation of Piwi-interacting RNAs at their 3' ends. Genes Dev 2007; 21:1603-8.
40. Kirino Y, Mourelatos Z. The mouse homolog of HEN1 is a potential methylase for Piwi-interacting RNAs. RNA 2007; 13:1397-1401.
41. Okamura K, Liu N, Lai EC. Distinct mechanisms for microRNA strand selection by Drosophila argonautes. Mol Cell 2009; 36:431-44.

42. Czech B, Zhou R, Erlich Y et al. Hierarchical rules for argonaute loading in Drosophila. Mol Cell 2009;
    36:445-456.
43. Ghildiyal M, Xu J, Seitz H et al. Sorting of Drosophila small silencing RNAs partitions microRNA* strands
    into the RNA interference pathway. RNA 2010; 16:43-56.
44. Li J, Yang Z, Yu B et al. Methylation protects miRNAs and siRNAs from a 3′-end uridylation activity in
    Arabidopsis. Curr Biol 2005; 15:1501-7.
45. Kurth HM, Mochizuki K. 2′-O-methylation stabilizes piwi-associated small RNAs and ensures DNA
    elimination in Tetrahymena. RNA 2009; 15:675-85.
46. Liu Y, Ye X, Jiang F et al. C3PO, an endoribonuclease that promotes RNAi by facilitating RISC activation.
    Science 2009; 325:750-3.
47. Wang B, Li S, Qi HH et al. Distinct passenger strand and mRNA cleavage activities of human argonaute
    proteins. Nat Struct Mol Biol 2009; 16:1259-66.
48. Kim K, Lee YS, Carthew RW. Conversion of preRISC to holo-RISC by Ago2 during assembly of RNAi
    complexes. RNA 2007; 13:22-29.
49. Schwarz DS, Hutvagner G, Du T et al. Asymmetry in the assembly of the RNAi enzyme complex. Cell
    2003; 115:199-208.
50. Khvorova A, Reynolds A, Jayasena SD. Functional siRNAs and miRNAs exhibit strand bias. Cell 2003;
    115:209-16.
51. Eamens AL, Smith NA, Curtin SJ et al. The Arabidopsis thaliana double-stranded RNA binding protein
    DRB1 directs guide strand selection from microRNA duplexes. RNA 2009; 15:2219-35.
52. Ro S, Park C, Young D et al. Tissue-dependent paired expression of miRNAs. Nucleic Acids Res 2007;
    35:5944-53.
53. Wei J-X, Yang J, Sun J-F et al. Both strands of siRNA have potential to guide post-transcriptional gene
    silencing in mammalian cells. PLoS ONE 2009; 4:e5382.
54. Vaucheret H. Plant ARGONAUTES. Trends Plant Sci 2008; 13:350-8.
55. Li CF, Pontes O, El-Shami M et al. An ARGONAUTE4-containing nuclear processing center colocalized
    with Cajal bodies in Arabidopsis thaliana. Cell 2006; 126:93-106.
56. Mi S, Cai T, Hu Y et al. Sorting of small RNAs into Arabidopsis argonaute complexes is directed by the
    5′ terminal nucleotide. Cell 2008; 133:116-27.
57. Takeda A, Iwasaki S, Watanabe T et al. The mechanism selecting the guide strand from small RNA duplexes
    is different among argonaute proteins. Plant Cell Physiol 2008; 49:493-500.
58. Qi Y, Denli AM, Hannon GJ. Biochemical specialization within Arabidopsis RNA silencing pathways.
    Mol Cell 2005; 19:421-28.
59. Montgomery TA, Howell MD, Cuperus JT et al. Specificity of ARGONAUTE7-miR390 interaction and
    dual functionality in TAS3 trans-acting siRNA formation. Cell 2008; 133:128-41.
60. Park MY, Wu G, Gonzalez-Sulser A et al. Nuclear processing and export of microRNAs in Arabidopsis.
    Proc Natl Acad Sci USA 2005; 102:3691-96.
61. Han MH, Goud S, Song L et al. The Arabidopsis double-stranded RNA-binding protein HYL1 plays a role
    in microRNA-mediated gene regulation. Proc Natl Acad Sci USA 2004; 101:1093-98.
62. Weinmann L, Hock J, Ivacevic T et al. Importin 8 is a gene silencing factor that targets argonaute proteins
    to distinct mRNAs. Cell 2009; 136:496-507.
63. Castanotto D, Lingeman R, Riggs AD et al. CRM1 mediates nuclear-cytoplasmic shuttling of mature
    microRNAs. Proc Natl Acad. Sci USA 2009; 106:21655-59.
64. Ramachandran V, Chen X. Degradation of microRNAs by a family of exoribonucleases in Arabidopsis.
    Science 2008; 321:1490-2.
65. Kennedy S, Wang D, Ruvkun G. A conserved siRNA-degrading RNase negatively regulates RNA interference
    in C. elegans. Nature 2004; 427:645-9.
66. Iida T, Kawaguchi R, Nakayama J. Conserved ribonuclease, Eri1, negatively regulates heterochromatin
    assembly in fission yeast. Curr Biol 2006; 16:1459-64.
67. Duchaine TF, Wohlschlegel JA, Kennedy S et al. Functional proteomics reveals the biochemical niche of
    C. elegans DCR-1 in multiple small-RNA-mediated pathways. Cell 2006; 124:343-54.
68. Gabel HW, Ruvkun G. The exonuclease ERI-1 has a conserved dual role in 5.8S rRNA processing and
    RNAi. Nat Struct Mol Biol 2008; 15:531-33.
69. Chatterjee S, Großhans H. Active turnover modulates mature microRNA activity in Caenorhabditis elegans.
    Nature 2009; 461:546-9.
70. Mullen TE, Marzluff WF. Degradation of histone mRNA requires oligouridylation followed by decapping
    and simultaneous degradation of the mRNA both 5′ to 3′ and 3′ to 5′. Genes Dev 2008; 22:50-65.
71. Wilusz CJ, Wilusz J. New ways to meet your (3′) end—oligouridylation as a step on the path to destruction.
    Genes Dev 2008; 22:1-7.

72. Zimmer SL, Fei Z, Stern DB. Genome-based analysis of Chlamydomonas reinhardtii exoribonucleases and poly(A) polymerases predicts unexpected organellar and exosomal features. Genetics 2008; 179:125-36.
73. Gy I, Gasciolli V, Lauressergues D et al. Arabidopsis FIERY1, XRN2 and XRN3 are endogenous RNA silencing suppressors. Plant Cell 2007; 19:3451-61.
74. Ibrahim F, Rohr J, Jeong WJ et al. Untemplated oligoadenylation promotes degradation of RISC-cleaved transcripts. Science 2006; 314:1893.
75. Schmid M, Jensen TH. The exosome: a multipurpose RNA-decay machine. Trends Biochem Sci 2008; 33:501-510.
76. Belostotsky D. Exosome complex and pervasive transcription in eukaryotic genomes. Curr Opin Cell Biol 2009; 21:352-8.
77. Zuo Y, Deutscher MP. Exoribonuclease superfamilies: structural analysis and phylogenetic distribution. Nucleic Acids Res 2001; 29:1017-26.
78. Nagaike T, Suzuki T, Katoh T et al. Human mitochondrial mRNAs are stabilized with polyadenylation regulated by mitochondria-specific poly(A) polymerase and polynucleotide phosphorylase. J Biol Chem 2005; 280:19721-7.
79. Rissland OS, Mikulaslova A, Norbury CJ. Efficient RNA polyuridylation by noncanonical poly(A) polymerases. Mol Cell Biol 2007; 27:3612-24.
80. Katoh T, Sakaguchi Y, Miyauchi K et al. Selective stabilization of mammalian microRNAs by 3′ adenylation mediated by the cytoplasmic poly(A) polymerase GLD-2. Genes Dev 2009; 23:433-8.
81. Lu S, Sun Y-H, Chiang VL. Adenylation of plant miRNAs. Nucleic Acids Res 2009; 37:1878-85.
82. Lange H, Holec S, Cognat V et al. Degradation of a polyadenylated rRNA maturation by-product involves one of the three RRP6-like proteins in Arabidopsis thaliana. Mol Cell Biol 2008; 28:3038-44.
83. Zhou R, Hotta I, Denli AM et al. Comparative analysis of Argonaute-dependent small RNA pathways in Drosophila. Mol Cell 2008; 32:592-9.
84. van Wolfswinkel JC, Claycomb JM, Batista PJ et al. CDE-1 affects chromosome segregation through uridylation of CSR-1-bound siRNAs. Cell 2009; 139:135-48.
85. Felice KM, Salzman DW, Shubert-Coleman J et al. The 5′ terminal uracil of let-7a is critical for the recruitment of mRNA to Argonaute2. Biochem J 2009; 422:329-41.
86. Ebhardt HA, Tsang HH, Dai DC et al. Meta-analysis of small RNA-sequencing errors reveals ubiquitous post-transcriptional RNA modifications. Nucleic Acids Res 2009; 37:2461-70.
87. Haley B, Zamore PD. Kinetic analysis of the RNAi enzyme complex. Nat Struct Mol Biol 2004; 11:599-606.
88. Jinek M, Doudna JA. A three-dimensional view of the molecular machinery of RNA interference. Nature 2009; 457:405-12.
89. Orban TI, Izaurralde E. Decay of mRNAs targeted by RISC requires XRN1, the Ski complex and the exosome. RNA 2005; 11:459-69.
90. Souret FF, Kastenmayer JP, Green PJ. AtXRN4 degrades mRNA in Arabidopsis and its substrates include selected miRNA targets. Mol Cell 2004; 15:173-83.
91. Kai ZS, Pasquinelli AE. microRNA assassins: factors that regulate the disappearance of miRNAs. Nat Struct Mol Biol 2010; 17:5-10.
92. Hwang H-W, Wentzel EA, Mendell JT. A hexanucleotide element directs microRNA nuclear import. Science 2007; 315:97-100.
93. Jones MR, Quinton LJ, Blahna MT et al. Zcchc11-dependent uridylation of microRNA directs cytokine expression. Nat Cell Biol 2009; 11:1157-63.
94. LaCava J, Houseley J, Saveanu C et al. RNA degradation by the exosome is promoted by a nuclear polyadenylation complex. Cell 2005; 121:713-24.
95. Wyers F, Rougemaille M, Badis G et al. Cryptic pol II transcripts are degraded by a nuclear quality control pathway involving a new poly(A) polymerase. Cell 2005; 121:725-37.
96. Vanacova S, Wolf J, Martin G et al. A new yeast poly(A) polymerase complex involved in RNA quality control. PLoS Biol 2005; 3:e189.
97. Ma J-B, Ye K, Patel DJ. Structural basis for overhang-specific small interfering RNA recognition by the PAZ domain. Nature 2004; 429:318-22.

CHAPTER 12

MicroRNases AND THE REGULATED DEGRADATION OF MATURE ANIMAL miRNAs

Helge Großhans* and Saibal Chatterjee

Abstract: microRNAs (miRNAs) are small noncoding RNAs that regulate numerous target mRNAs through an antisense mechanism. Initially thought to be very stable with half-lives on the order of days, mature miRNAs have recently been shown to be subject to degradation by 'microRNases' (miRNases) in plants (the small RNA degrading nucleases, SDN) and animals (exoribonuclease 2/XRN-2/XRN2). Interference with these miRNA turnover pathways causes excess miRNA activity, consistent with an important contribution to miRNA homeostasis. Moreover, it is now emerging that long half-lives are not an invariant feature of miRNAs but that marked differences exist in the stabilities of individual miRNAs and that cellular states can further determine miRNA turnover rates. Although the means of regulation are still largely unclear, biochemical data suggest that target mRNA-binding can stabilize miRNAs within their Argonaute (AGO) effector complexes, providing one possible mechanism that may control miRNA half-lives. We will summarize here what is known about miRNA turnover in animals and how recent discoveries have established a new dynamic of miRNA-mediated gene regulation. We will highlight some of the open questions in this emerging area of research.

INTRODUCTION

microRNAs (miRNAs) are small regulatory RNAs, about 22 nucleotides in length, that bind to partially complementary messenger RNAs (mRNAs) and repress them translationally or by transcript degradation.[1] The discovery, about a decade ago, that miRNAs modulate the expression of a substantial fraction of animal genes[2-4] came as a major surprise, as this extensive post-transcriptional control of gene expression necessitated major revisions to our view of how genes are regulated.

*Corresponding Author: Helge Großhans—Friedrich Miescher Institute for Biomedical Research, PO Box 2543, CH-4002 Basel, Switzerland. Email: helge.grosshans@fmi.ch

Regulation of microRNAs, edited by Helge Großhans.
©2010 Landes Bioscience and Springer Science+Business Media.

Over the past few years, we have now learned that miRNA expression itself is also heavily regulated at the post-transcriptional level,[5] adding yet another layer of complexity to gene regulation. However, regulation appeared to be largely restricted to modulation of miRNA biogenesis, as mature miRNAs were considered exceptionally stable,[6] wedded for life to their Argonaute (AGO) effector protein. Evidence for this was derived from a few observations of miRNA turnover rates in cultured cells or in tissues,[7-9] as well as the typically slow and/or incomplete response of mature miRNA levels to alterations in miRNA biogenesis rates (Fig. 1).[6,10-14] Consistent with this view appeared the observation that miRNAs were also extremely stable in fixed tissue samples, where they could be detected for years[15]—unlike the longer mRNAs, which are substantially degraded under these conditions. However, there were already hints that not all miRNAs were created equal so that some miRNAs might be less stable, at least under certain conditions, than expected.[8] Indeed, given the dynamic of miRNA expression during development,[16-18] with some miRNAs being up-, others down-regulated, it appeared unlikely that the long half-lives reported for some miRNAs, measuring days or more, could hold true under all conditions, for all miRNAs.

A first observation hinting at the possibility of active and regulated miRNA turnover was made in plants, namely in the thale cress *Arabidopsis thaliana*: Xuemei Chen and coworkers identified a family of small RNA degrading nucleases (SDNs) that could degrade synthetic miRNAs in vitro and whose joint depletion elevated mature miRNA levels in vivo[19] (see chapter by Cerutti and Ibrahim for details). Subsequently, our lab identified a different nuclease, XRN-2, as a mediator of miRNA degradation in the roundworm *C. elegans*.[20] These RNases belong not only to distinct protein families, they also act through distinct mechanisms, SDNs being 5'→3' exonucleases, but XRN-2 a 5'→3' exonuclease. Intriguingly, however, the yeast SDN homologues Rex1p through Rex4p are involved in ribosomal RNA (rRNA) biogenesis, as is yeast Rat1p, the XRN-2 orthologue (see also separate section below). It was noted previously[21] that Drosha, the RNase that processes pri-miRNAs into pre-miRNAs,[10] and its associated helicases DDX5 and DDX17 (also known as P68 and P72),[22] are also all involved in rRNA biogenesis.[23,24] There thus appears to be a theme of overlap between miRNA and rRNA pathways, although the rationale for this, if any, is still unclear.

We will discuss here what is known about miRNA turnover in animals and its potential effects on gene expression. We will particularly focus on the XRN-2 miRNase in *C. elegans*, but additionally highlight studies hinting at a wide-spread occurrence of regulated miRNA degradation as a means to achieve dynamic gene regulation. We will also discuss some of the open questions in the field that will need to be addressed in future studies.

microRNA BIOGENESIS AND FUNCTION

As detailed reviews on miRNA biogenesis and function are provided in the chapter by Ketting and elsewhere,[25] this section will give only a brief summary of the most relevant points. Briefly, miRNAs are transcribed by RNA polymerase II as long, primary miRNAs (pri-miRNAs) that are capped and polyadenylated.[26-28] In addition to these independent, 'intergenic' pri-miRNAs, a substantial fraction of miRNAs, particularly in mammals, appears to be cotranscribed with 'host' pre-mRNAs, in the introns of which they reside.[29,30] The RNase III-type enzyme Drosha processes both intergenic and intronic

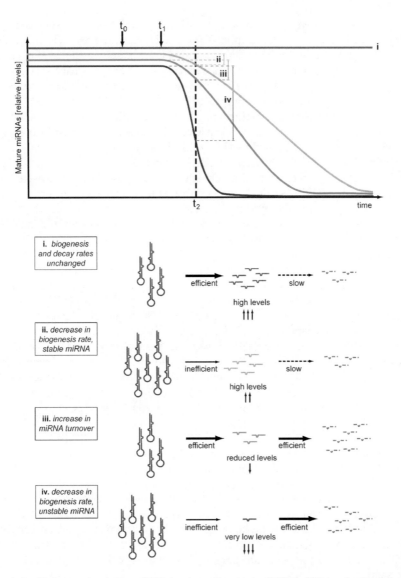

Figure 1. Rapid changes in mature miRNA levels require short miRNA half-lives. Accumulation of a mature, stable miRNA is at equilibrium at t_0. At t_1, miRNA decay and/or biogenesis rates are altered. The graph (*top*) depicts the changes (colored double-arrows) that will be observed at t_2, a time point shortly after t_1 (i.e., $t_2-t_1 \ll t_{\frac{1}{2};0}$ where $t_{\frac{1}{2};0}$ is the half-life of the miRNA at t_0).
i) Both biogenesis and decay rates remain unaltered, thus leaving mature miRNA levels unchanged at t_1.
ii) Biogenesis rate is reduced but miRNA decay remains slow, thus causing only a modest decrease in mature miRNA levels at t_2. iii) Biogenesis rate is unchanged, but the decay rate increased, causing a notable decrease in mature miRNA levels at t_2. iv) miRNA decay is increased and the biogenesis rate is decreased, causing a substantial decrease in mature miRNA levels at t_2.
Note that under the conditions depicted here, a decline in miRNA biogenesis rate has little effect on the levels of stable mature miRNAs at t_2 (compare i and ii). Conversely, an identical decrease in miRNA biogenesis causes a much stronger decline in the levels of unstable mature miRNAs (i.e., $t_2-t_1 \geq t_{\frac{1}{2};0}$) (compare iii and iv). Also note that in these examples miRNA biogenesis is assumed to be impaired at the pre-miRNA processing step, leading to pre-miRNA accumulation. No such accumulation would be observed if other steps of miRNA biogenesis such as transcription, or pri-miRNA processing were impaired.

pri-miRNAs, releasing by cleavage the so-called precursor miRNA (pre-miRNA),[10] a processing intermediate that is characterized by a stem-loop structure of approximately 70 nucleotides. The pre-miRNA is exported into the cytoplasm,[13,31,32] where another RNase III enzyme, Dicer, processes it into a duplex RNA consisting of miRNA guide and passenger strands, also known as miR and miR*.[33-35] Incorporation of the miRNA guide strand into AGO yields an miRNA-induced silencing complex (miRISC), whereas the passenger strand is discarded.[25] Within miRISC, the miRNA serves as an 'address label' that targets the complex to partially complementary sequences in mRNAs that will be silenced by miRISC binding.[1] Although the precise composition of miRISC is controversial, recent data from several systems have provided strong evidence for GW182 proteins, also known as TNRC6 in humans and AIN-1, AIN-2 in *C. elegans*, as important components and downstream effectors of AGO.[36] The mechanism of miRNA silencing has also been hotly debated, with proposed mechanisms involving translational repression at the initiation or elongation level, mRNA degradation with or without deadenylation and cotranslational protein degradation. However, recently there has been a convergence on mRNA deadenylation and degradation and repression of translation initiation as the major mechanisms.[37]

The latter mechanism is of particular interest, as it permits reversible repression of miRNA targets,[9] as discussed in the chapter by Meisner and Filipowicz. By contrast, a degradation mechanism would imply permanent target silencing. Moreover, such an mRNA clearance mechanism would ultimately cause an accumulation of 'unemployed' miRISC lacking targets. This is unless there is either a limited amount of miRISC programmed with a specific miRNA that silences an excess of cognate targets (which would permit continuous miRISC reuse), or active degradation of miRISC and/or its associated miRNA when they are devoid of targets. With the discovery of 'miRNases', the latter scenario now appears to be a distinct possibility and as we will discuss below, we have demonstrated that miRNAs can indeed be dislodged from target-free miRISC to become subject to degradation by XRN-2.[20]

XRN-2, A MULTIFUNCTIONAL EXORIBONUCLEASE

The 5'→3' exonuclease XRN-2 is conserved across eukaryotes and also known as exoribonuclease 2, XRN2 and, in yeast, variously Rat1p, Hke1p or Tap1p. We will use here XRN-2, except when referring explicitly to the yeast protein, for which we will use Rat1p, both for *Saccharomyces cerevisiae* and *Schizosaccaromyces pombe*.

It is in yeast that Rat1p/XRN-2 has been particularly well characterized, ironically a eukaryote that does not have an miRNA pathway. Rat1p is required for degradation of pre-mRNAs[38] and noncoding telomeric RNAs,[39] the 5' processing of ribosomal and small nuclear RNAs[40-42] and transcriptional termination.[43] A role in transcriptional termination has also been reported for mammalian XRN-2.[44] However, transcriptional termination activity might be unrelated to exonucleolytic RNA cleavage by Rat1p/XRN-2 and the enzyme appears largely dispensable for correct termination.[45,46]

In addition to its role in miRNA turnover, which we will discuss below, XRN-2 also functions during miRNA biogenesis. However, this activity does not appear to affect mature miRNA levels, but instead is needed to clear away potentially harmful byproducts of miRNA processing. Thus, it has been shown that for intronic miRNAs, cleavage by Drosha can occur cotranscriptionally, prior to splicing.[47] Clearance of the remaining intron sequence involves XRN-2, together with the exosome, and depletion of XRN-2 impairs splicing of the host gene. XRN-2 also promotes transcriptional termination at intergenic miRNA loci, but it appears that this function might have no direct bearing on levels or activity of the miRNAs.[48] Finally, in *Arabidopsis*, where XRN2, XRN3 and XRN4 are three paralogues whose closest animal homologue is XRN-2, XRN2 and XRN3 are involved in degradation of the pre-miRNA loop following processing of the pre-miRNA by Dicer.[49] Again, this scavenger-type function of XRN-2-related enzymes did not affect mature miRNA levels.

Many of the functions of Rat1p overlap with those of its paralogue Xrn1p, although the two proteins differ in their preferential intracellular localizations, with Rat1p being nuclear and Xrn1p cytoplasmic.[50] However, Xrn1p contributes to nuclear functions such as snoRNA and rRNA processing[40-42,51] or cotranscriptional RNA degradation,[45] whereas *RAT1*, at least when over-expressed, can suppress loss of *XRN1*[52] (although the reverse does indeed require targeting of Xrn1p to the nucleus).[50,53] Thus, the clear separation into nuclear and cytoplasmic exonucleases appears to be an oversimplification based on steady-state localization of the involved enzymes. It hardly requires a stretch of imagination to propose that the actual localizations might be more dynamic and/or that minor pools of Rat1p in the cytoplasm and Xrn1p in the nucleus could exist[54] that can sustain some but not all functions of the respective paralogue.

As in other animals, both Rat1p (XRN-2) and Xrn1p (XRN-1) orthologues exist in *C. elegans*, but so far a function in miRNA turnover has only been established for XRN-2, although a function of XRN-1 remains possible.[20]

DEGRADATION OF *C. ELEGANS* miRNAs BY XRN-2

We have described[20] the identification of XRN-2 as a nuclease that altered miRNA levels and activity in *C. elegans* by degrading mature miRNAs upon their dislodging from the Argonaute effector protein ("AGO"; the paralogous ALG-1 and ALG-2 proteins in *C. elegans*).[34] Specifically, incubation with *C. elegans* larval lysate caused degradation of synthetic, radiolabelled miRNAs and this degradation was impaired when the lysate had been depleted for XRN-2. In a more physiological scenario, we confirmed that XRN-2 activity was not limited to 'naked' RNA, but also occurred on protein-covered miRNA, i.e., the kind of substrate that is likely to be encountered in the cell. Thus, when pre-miRNA was added to *C. elegans* lysate, it was processed by Dicer to yield the mature miRNA, which was incorporated into AGO. However, in the presence of XRN-2, this mature miRNA would be released again from AGO and degraded by XRN-2 (Fig. 2). How the release step itself is achieved, is currently unknown. Although XRN-2 is important for this function in vivo (see below), it appears to play only an auxiliary role in vitro. Indeed, given that the 5' end of the mature miRNA is thought to be buried within the AGO Mid domain,[55] and thus not accessible to XRN-2, it appears unlikely that XRN-2 would be the primary release factor, at least in a model where its release activity would be directly coupled to exonucleolytic decay of the miRNA.

We confirmed that our assays recapitulated a true physiological function of XRN-2, by examining the effects of RNAi-mediated depletion of *xrn-2* on miRNA levels in vivo.[20] We found that, relative to a mock RNAi control, levels of different mature miRNAs, but not of their pre- or pri-miRNA precursors were elevated (Fig. 3). Importantly, this elevation also coincided with enhanced miRNA activity, as demonstrated by the

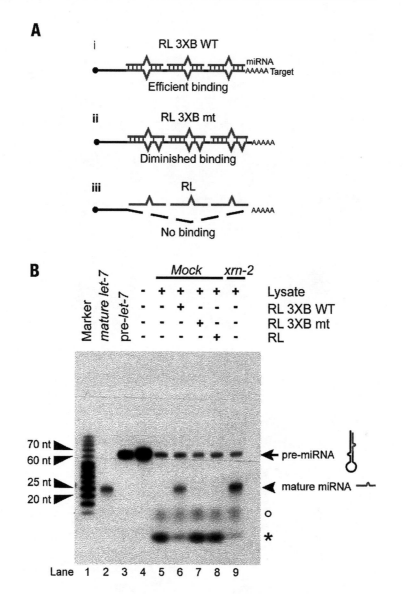

Figure 2. Accumulation of a mature miRNA from a pre-miRNA requires protection through a target mRNA or reduced XRN-2 levels in a *C. elegans* lysate. (Adapted from ref. 20.) A) Schematic depiction of the in vitro RNA transcripts used in (B) (not to scale). The Renilla luciferase (RL) reporter mRNAs contain 3'UTRs with (i) three functional *let-7* binding sites and (ii) three mutated sites or (iii) lack a 3'UTR. B) Pre-*let-7* turnover using *N2* (*C. elegans* wild-type) lysate in the absence or presence of three different mRNAs as indicated and schematically shown in (A). Radiolabelled substrate and product are indicated. Note that all lysates equally reduce pre-*let-7* levels relative to input (compare lanes 5-9 to lane 4), but that mature *let-7* only accumulates if either an mRNA with functional miRNA-binding sites is present (lane 6) or *xrn-2* is depleted by RNAi (lane 9). Conversely, the mononucleotide decay product (asterisk) is depleted under these conditions, whereas an independent decay product (circle), presumably the partially degraded loop that dicing of the pre-miRNA releases, remains unchanged. Labels indicate lysates prepared from animals exposed to mock RNAi and *xrn-2(RNAi)*, respectively.

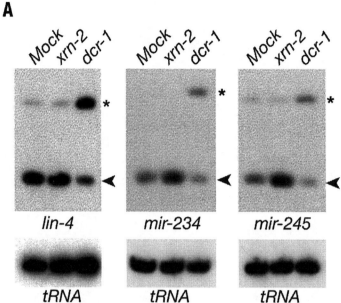

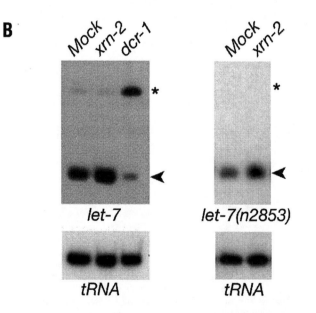

Figure 3. Depletion of *xrn-2* causes accumulation of mature miRNAs but not their precursors in vivo. A) Northern blot analysis reveals that RNAi-mediated depletion of *xrn-2* elevates the levels of mature *mir-234* and *mir-245* relative to mock RNAi, but barely affects *lin-4* levels (arrowheads), possibly indicating substrate specificity or technical limitations (see main text for discussion). B) Elevation is moderate (~2-fold) for mature *let-7* in wild-type animals, but more substantial for the mutant *let-7(n2853)* variant, the levels of which are reduced relative to wild-type animals. A,B) Pre-miRNA levels (asterisks) are unchanged for any miRNA, whereas disruption of pre-miRNA processing by depletion of Dicer (*dcr-1*) results in elevated pre-miRNA levels. Figure in part adapted from reference 20.

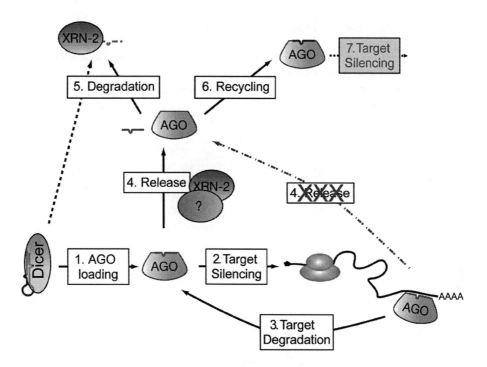

Figure 4. A model of XRN-2 affecting miRNA homeostasis and AGO recycling. Following processing of the pre-miRNA by Dicer and unwinding of the passenger:guide duplex, (1) mature miRNA is loaded into Argonaute (AGO) to (2) silence its targets. If no target is available or (3) the target has been degraded by the miRNA:AGO complex, (4) the miRNA can be released from AGO for (5) degradation by XRN-2. This may permit (6) recycling of AGO and loading with a different miRNA, which (7) can silence other targets. Release of the miRNA from AGO is promoted by XRN-2, but appears to involve additional factors, yet to be identified. Note that the presence of target mRNA prevents release of the miRNA from AGO (right dashed line). Left dotted line: a fraction of mature miRNA might also be degraded directly, without prior incorporation into RISC. Modified from reference 20.

increased repression of *let-7* miRNA targets in a *let-7(n2853)* mutant background. In this mutant, a single point mutation in the mature miRNA causes, for unknown reasons, a decrease in mature *let-7* levels and a desilencing of *let-7* targets and both of these effects are reversed upon *xrn-2* depletion. Intriguingly, the increase in mature *let-7* levels caused by *xrn-2(RNAi)* was fairly modest (Fig. 3)—approximately three-fold relative to the mock RNAi control—yet this sufficed not only to restore repression of the *let-7* targets, but also to suppress the lethality, as well as additional developmental phenotypes caused by reduced *let-7* activity. This observation thus provides evidence for the importance of precise and faithful regulation of miRNA levels during animal development.

Since activity of the miRNA requires its incorporation into the miRISC effector complex, specifically, its binding to AGO, we can further conclude that in vivo, XRN-2 is indeed important for miRNA release from AGO, whereas it appears to play only a facilitating role in vitro (Fig. 4). One speculative scenario to reconcile these differences would be the existence of a limited pool of dedicated miRNA release factor, the recycling of which would require degradation of the unloaded miRNA by XRN-2.

IS miRNA TURNOVER A SUBSTRATE-SPECIFIC EVENT?

Exoribonucleases of the XRN1/XRN2 family prefer 5'-monophosphorylated, single-stranded and unstructured RNAs as substrates, whereas 5' triphosphates, 5' mRNA caps, or RNA structures severely impair or even entirely abrogate activity in vitro.[54] Little additional specificity is known to exist and we confirmed that four different synthetic mature miRNAs were similarly degraded in *C. elegans* lysate.

However, it remains a strong possibility that additional substrate specificity could exist in vivo, conferred through cofactors or through distinct spatial organization of miRNA turnover. In yeast, for instance, Rat1p functions with a cofactor, the Rat1-interacting protein 1, Rai1p, which, in *Schizosaccharomyces pombe*, is required for processivity of Rat1p-mediated RNA decay.[56] More intriguingly, Rai1p possesses pyrophosphohydrolase activity, which hydrolyzes 5' triphosphate ends on RNAs to yield 5'-monophosphorylated Rat1p substrates.[56] However, structural and interaction studies suggest that the higher eukaryote Rai1p orthologue Dom3Z does not physically associate with XRN-2,[56,57] leaving it unclear if and which cofactors act with non-yeast XRN-2.

Superficially consistent with the notion of increased substrate specificity under physiological conditions, we found when we examined the effect of *xrn-2* depletion on 12 different miRNAs in vivo[20] that the levels of nine were at least twofold elevated, whereas the other three revealed only minor accumulation (Fig. 3). However, it is unclear whether this reflects true substrate specificity, as all of the weakly affected miRNAs, such as *lin-4*, tended to be expressed already early during *C. elegans* development, in the first larval (L1) stage.[16] As our experiments involved exposure of synchronized L1 stage larvae to *xrn-2(RNAi)* and subsequent harvesting of the animals at the L4 stage, it appears likely that during L1 (and perhaps even some later stages), sufficient XRN-2 would still have been available to shape the expression patterns of these miRNAs. Clearly, more extensive miRNA profiling will be needed to determine whether XRN-2 does indeed exhibit specificity with regard to the miRNAs that it degrades.

A FUNCTION OF XRN-2 IN miRNA TURNOVER BEYOND *C. ELEGANS*?

At this point, it is not known whether XRN-2 also effects degradation of miRNAs in other organisms, although its high level of conservation would support this notion. Intriguingly, however, a recent report links *XRN2* to lung cancer in mice and humans.[58] Specifically, a genome-wide association study aimed at identifying gene loci relevant to spontaneous (non cigarette smoke-induced) lung cancer in mice revealed a strong association between lung cancer development and single nucleotide polymorphisms (SNPs) located near the *XRN2* locus in mice. Further studies revealed that this correlation extended to altered *XRN2* expression levels, with animals more at risk for lung cancer expressing higher levels of *XRN2*. Moreover, in vitro (in cell culture), over-expression of *XRN2* in human Beas-2B lung epithelial cells increased their proliferation and reduced the expression of differentiation markers. Finally, an association between lung cancer and SNPs in *XRN2* was also confirmed for human lung cancer. These findings are of particular interest when considering the number of connections between lung cancer and miRNA expression, in particular the fact that the human *let-7* miRNAs act as tumor suppressors in the lung, where they repress *HMGA2*, *RAS* and other proto-oncogenes.[59] It will be of substantial interest to learn whether this novel link between XRN2 and lung cancer does indeed reflect a function of mammalian XRN2 in miRNA degradation, perhaps even geared towards specific miRNAs such as the *let-7* family.

HALF-LIVES OF miRNAs—NOT ALL miRNAs ARE EQUAL

As hinted earlier in this chapter, it has long been assumed that mature miRNAs are highly stable molecules.[6] Half-lives of 14 hours and more[7-9,14] appeared to argue against the presence of a vigorous miRNA turnover activity, as did the stability of miRNAs in fixed tissue samples[15]—although the latter could hardly be considered to reflect a physiological situation. However, in vitro pre-miRNA processing reactions using Drosophila lysates revealed very inefficient accumulation of mature miRNA, although the pre-miRNA substrate was readily consumed.[33] One possible explanation for this observation, not further investigated at that time, was that mature miRNA was made, but subsequently efficiently degraded. Even within cells, not all miRNAs appeared to behave the same:[8] Whereas a synthetic miR-29a transfected into HeLa cells (above a background of endogenous miR-29a and miR-29b, produced from the same pri-miRNA) had a half-life of ~14 hours, its sequence-related 'sister' miR-29b, had a half-life of only 4 hours. Moreover, when cells were arrested in mitosis, both miRNAs were stabilized to half-lives exceeding 12 hours, reflecting the fact that the endogenous miR-29b accumulates specifically during mitosis. Thus, it appeared that both elements within the miRNA and cellular states could affect miRNA stability, at least at the elevated levels achieved through transfection.

A recent report revealed that the accumulation of additional endogenous miRNAs (which did not include miR-29b) also varied substantially across cell cycle stages in HeLa cells.[60] Since an entire cell cycle under these conditions takes some 16 hours, these findings would suggest that also for some endogenous miRNAs, decay has to occur much more rapidly than suggested by half-lives of 14 or more hours (Fig. 1), although precise half-life measurements are yet to be reported.

Direct evidence for the notion that endogenous miRNAs can decay rapidly comes from a study on postmortem human brain tissues and primary human neuron cultures,[61] which revealed half-lives ranging from less than an hour to under four hours for miR-9, miR-125b, miR-132, miR-146a and miR-183. As other miRNAs were not investigated, it is unclear whether these selected miRNAs have unusually short half-lives or whether miRNA decay in brain is generally accelerated.

Regardless of these possibilities, the notion that some or all neuronal miRNAs could have short half-lives is particularly intriguing when considering the importance of translational control at synapses and, more specifically, the observation that target silencing by miR-134 at synapses can be readily reversed, albeit through an unknown mechanism.[62] Intriguingly then, it was found in *Aplysia californica* (sea slug) neurons that the levels of miR-124, a particularly abundant, brain-specific miRNA, were rapidly reduced upon treatment of sensory neurons in cell culture with the neurotransmitter serotonin.[63] The precise kinetics of miR-124 depletion are unclear, as treatment involves five pulses of five minutes' treatment with serotonin with pulses being spaced 20 minutes apart (i.e., 105 min for the entire procedure). However, a two-fold reduction was already apparent at one hour after the last pulse, further extending to a maximum of threefold reduction two hours after the last pulse, before miR-124 amounts slowly returned to baseline levels. When MAP kinase signaling was abrogated through use of the MAPK inhibitor U0126, serotonin failed to induce a reduction of miR-124 levels, implicating this cell signaling pathway in the regulated clearance of miR-124, although the relevant effector proteins remain to be identified.[63]

The reduction in miR-124 levels upon serotonin treatment, although modest, could be functionally relevant: One of the targets of miR-124 is the transcriptional activator CREB1 (Cyclic AMP-responsive element-binding protein 1), the overexpression of which enhances long-term facilitation, i.e., synaptic activity, and inhibition of miR-124 was found to phenocopy this effect.[63]

At this point, it is not completely clear whether it is truly the decay of miR-124 or its biogenesis rate that is altered by serotonin exposure. Precursor levels of miR-124 were reported to remain stable, suggesting that serotonin-treatment indeed increased mature miRNA turnover rate.[63] However, as the PCR-assay that was used to detect miR-124 precursors cannot distinguish between pri- and pre-miRNA, the relative abundance of which is also unknown, it remains equally possible that the decrease in miR-124 levels reflects a selective reduction in processing of either precursor against a background of constitutively high turnover of miR-124 (Fig. 1).

In a different system, immortalized MCF10A cells that contain a fusion of the estrogen receptor (ER) domain with the Src oncoprotein, rapid, possibly regulated turnover of mature miRNAs has also been observed. Treatment of these cells with tamoxifen activates Src and causes cellular transformation. Under these conditions, *let-7* miRNAs are lost with an approximate half-life of four hours and a 20-fold depletion occurs by 36 hours after tamoxifen treatment.[64] Based on the rapidity of this effect, Iliopoulos and colleagues speculated that this observation reflected a regulated mechanism of RNA degradation.[64] However, it remains to be experimentally established that the regulation indeed occurs at the level of mature *let-7* stability. The fact that LIN28B, a known regulator of pre- and pri-*let-7* processing[65-71] (reviewed in the chapter by Lehrbach and Miska), is required for the effect, would rather seem to point to impaired *let-7* biogenesis. Nonetheless, this would still require constitutively high turnover of *let-7* in MCF10A cells, as a block in processing of miRNA precursors would fail to deplete stable mature miRNAs rapidly (Fig. 1).

Finally, selective stabilization of a specific miRNA, miR-122, has been reported to occur in mouse liver.[72] Mature miR-122 undergoes 3'-adenylation by the noncanonical cytoplasmic poly(A)-polymerase GLD2 and loss of GLD2 reduces steady-state levels and activity of the mature mir-122. Several question remain to be explored, namely, does GLD2 indeed affect the miR-122 half-life and does this depend on 3'-adenylation. If so, what is the mechanism? The notion that adenylation of miR-122 could interfere with a 3'→5' degradation pathway is clearly an attractive possibility that deserves to be tested.

In summary, although the investigation of miRNA half-lives has only begun, it has already become clear that miRNA stabilities can be quite variable. The short half-lives of some miRNAs further indicate that turnover has the potential to substantially affect miRNA steady-state levels and thus activity, providing a new dynamic to miRNA-mediated gene regulation. However, at this point little is known about the mechanisms that can destabilize miRNAs in any of the situations discussed here, let alone the factors that could direct degradation of specific miRNAs, in specific tissues or developmental situations.

TARGET AVAILABILITY AFFECTS RELEASE OF miRNAs FROM AGO AND THEIR SUBSEQUENT DEGRADATION

As evidence of regulation of miRNA turnover is scarce, little is known about the mechanisms that could be involved. However, our in vitro studies revealed one potential mechanism whereby the abundance of miRNA targets modulates accumulation of their cognate miRNAs within miRISC.[20]

Although a mature guide miRNA was readily degraded upon incubation in *C. elegans* larval lysate, this was not true for the guide:passenger duplex processing intermediate whose generation precedes guide strand incorporation into AGO.[20] Accordingly, when pre-miRNA was processed in a lysate depleted of XRN-2, the mature miRNA that accumulated was single-stranded. These findings were consistent with the requirement of XRN-2 for single-stranded RNA substrates,[54] and immediately raised the question whether the binding to target mRNA would prevent the miRNA from being degraded. Subsequent turnover studies in the presence of target transcripts confirmed this notion, revealing that co-incubation of pre-miRNA with a target mRNA, but not an equivalent mRNA lacking the miRNA target sites, led to accumulation of the mature miRNA (Fig. 2), incorporated into AGO.[20] The effect was not limited to *let-7*, a miRNA that is known to be highly regulated,[59] but also held true for a second miRNA that was investigated, *mir-237*.

Although regulation of miRNA degradation through target mRNA-binding still needs to be confirmed in vivo, its existence would make sense on theoretical grounds: Several studies have provided circumstantial evidence for AGO as a limiting factor in miRNA/siRNA pathways. For instance, when exogenous siRNA/miRNA duplexes are supplied to cells, endogenous miRNA targets tend to become upregulated,[73,74] consistent with competition for a limiting factor downstream of Dicer. Conversely, *AGO* overexpression results in an accumulation of mature miRNAs,[75] whereas *AGO* depletion reduces miRNA levels.[34,76] A pathway that permits alignment of programmed (that is, miRNA-bound) AGO levels with target mRNA levels, as well as recycling of AGO loaded with miRNAs not engaged with targets, would appear highly beneficial under such limiting AGO conditions. Interestingly, such a scenario would also imply that it is indeed the AGO release step that is particularly important, perhaps more so than the ultimate degradation of the mature miRNA.

We also note that proteins have been identified that prevent miRNAs from repressing specific targets[77,78] or, yet more dramatically, that reverse silencing of a target by a miRNA.[9] If such a 'target-less' programmed AGO could be recognized and its bound miRNA released and degraded, this would not only recycle AGO for re-use but also reinforce target de-silencing, preventing futile cycles of silencing and de-silencing. Finally, given that mRNA half-lives can be very short and extensively regulated in response to cellular or environmental cues, loss of miRNA protection through targets might also account for some of the changes in miRNA levels seen in various systems and discussed above. Intriguingly, in such a scenario reduced miRNA levels would be a consequence, not a cause of altered mRNA expression patterns.

CONCLUSION

So far, two classes of miRNases are known, SDN proteins in *Arabidopsis*[19] and XRN-2 in *C. elegans*.[20] Current data would suggest at face value that neither XRN2 in *Arabidopsis* nor SDN/Rex in *C. elegans* share these functions.[20,49] However, as usual, absence of evidence should not be taken as evidence of absence, i.e., although individual depletion of the *SDN* family members in *C. elegans* did not cause suppression of the *let-7(n2853)* mutation, it remains possible that these proteins could function redundantly, so that their individual depletion has no effect. Moreover, it is conceivable that they might have some substrate and/or tissue specificity that would prevent them from acting on *let-7*. Similarly, although *Arabidopsis* XRN2 and XRN3 (which, together with XRN4, form a family of paralogues that is most closely related to animal XRN2, not XRN1) appear to be involved in degradation of the pre-miRNA loop following processing of the pre-miRNA by Dicer, but do not seem to affect mature miRNA levels,[49] this does not rule out that they might also degrade some mature miRNAs, in some situations.

More generally, there is little reason to believe that other RNases could not be involved in miRNA degradation. For instance, miRNA accumulation upon codepletion of multiple SDN family members in Arabidopsis rarely exceeds three-fold the levels seen in control plants,[19] as does XRN-2 depletion in *C. elegans*.[20] Although incomplete depletion is likely to account for this at least in part, it seems reasonable to assume that additional RNases could support miRNA degradation. As some miRNAs accumulate more than others upon depletion of SDN or XRN-2, a miRNA-specific function would also appear possible for those two miRNases, although it will need to be ruled out that technical reasons account for the observed differences; e.g., time and space of expression of the miRNA vs depletion of the miRNase.

Work by Cerutti and colleagues[79] has recently revealed a role of an additional RNase, RRP6, in miRNA turnover. In the alga *Chlamydomonas reinhardtii*, miRNAs that lack the characteristic 2'-O-methylation of their 3' end that is typically found in plant and algal miRNAs can be uridylated by the nucleotidyltransferase MUT68. This 3' end uridylation in turn makes the miRNAs susceptible to degradation by RRP6 (see also the chapter by Cerutti and Ibrahim). Although this function appears to be related to miRNA quality control rather than regulation, it is tempting to speculate that lack or loss of the 2'-O-methyl modification could be favored under some circumstances, thus permitting regulated miRNA turnover.

We also note that although both the work of the Chen lab and ours showed the importance of degradation of functional miRNAs to maintain miRNA homeostasis, it remains to be shown if and how these processes are regulated. However, given that miRNA expression patterns can be very dynamic during development, there seems to be little question about the 'if' and it will be exciting to identify the cofactors and regulators of the miRNases that determine their activity at specific times or towards specific substrates.

ACKNOWLEDGEMENTS

Work in HG's lab is funded by the Novartis Research Foundation, the Swiss National Science Foundation , and an ERC Starting Independent Investigator Award ('miRTurn'). SC is supported by Marie Curie and EMBO longterm postdoctoral fellowships.

REFERENCES

1. Filipowicz W, Bhattacharyya SN, Sonenberg N. Mechanisms of post-transcriptional regulation by microRNAs: are the answers in sight? Nat Rev Genet 2008; 9:102-114.
2. Lau NC, Lim LP, Weinstein EG et al. An abundant class of tiny RNAs with probable regulatory roles in Caenorhabditis elegans. Science 2001; 294:858-862.
3. Lee RC, Ambros V. An extensive class of small RNAs in Caenorhabditis elegans. Science 2001; 294:862-864.
4. Lagos-Quintana M, Rauhut R, Lendeckel W et al. Identification of novel genes coding for small expressed RNAs. Science 2001; 294:853-858.
5. Ding XC, Weiler J, Großhans H. Regulating the regulators: mechanisms controlling the maturation of microRNAs. Trends Biotechnol 2009; 27:27-36.
6. Lee Y, Hur I, Park SY et al. The role of PACT in the RNA silencing pathway. EMBO J 2006; 25:522-532.
7. van Rooij E, Sutherland LB, Qi X et al. Control of stress-dependent cardiac growth and gene expression by a microRNA. Science 2007; 316:575-579.
8. Hwang HW, Wentzel EA, Mendell JT. A hexanucleotide element directs microRNA nuclear import. Science 2007; 315:97-100.
9. Bhattacharyya SN, Habermacher R, Martine U et al. Relief of microRNA-mediated translational repression in human cells subjected to stress. Cell 2006; 125:1111-1124.
10. Lee Y, Ahn C, Han J et al. The nuclear RNase III Drosha initiates microRNA processing. Nature 2003; 425:415-419.
11. Haase AD, Jaskiewicz L, Zhang H et al. TRBP, a regulator of cellular PKR and HIV-1 virus expression, interacts with Dicer and functions in RNA silencing. EMBO Rep 2005; 6:961-967.
12. Yi R, Qin Y, Macara IG et al. Exportin-5 mediates the nuclear export of pre-microRNAs and short hairpin RNAs. Genes Dev 2003; 17:3011-3016.
13. Lund E, Güttinger S, Calado A et al. Nuclear export of microRNA precursors. Science 2004; 303:95-98.
14. Gatfield D, Le Martelot G, Vejnar CE et al. Integration of microRNA miR-122 in hepatic circadian gene expression. Genes Dev 2009; 23:1313-1326.
15. Nelson PT, Baldwin DA, Scearce LM et al. Microarray-based, high-throughput gene expression profiling of microRNAs. Nat Methods 2004; 1:155-161.
16. Lim LP, Lau NC, Weinstein EG et al. The microRNAs of Caenorhabditis elegans. Genes Dev 2003; 17:991-1008.
17. Okamura K, Phillips MD, Tyler DM et al. The regulatory activity of microRNA* species has substantial influence on microRNA and 3' UTR evolution. Nat Struct Mol Biol 2008; 15:354-363.
18. Avril-Sassen S, Goldstein LD, Stingl J et al. Characterisation of microRNA expression in post-natal mouse mammary gland development. BMC Genomics 2009; 10:548.
19. Ramachandran V, Chen X. Degradation of microRNAs by a family of exoribonucleases in Arabidopsis. Science 2008; 321:1490-1492.
20. Chatterjee S, Großhans H. Active turnover modulates mature microRNA activity in Caenorhabditis elegans. Nature 2009; 461:546-549.
21. Gabel HW, Ruvkun G. The exonuclease ERI-1 has a conserved dual role in 5.8S rRNA processing and RNAi. Nat Struct Mol Biol 2008; 15:531-533.
22. Fukuda T, Yamagata K, Fujiyama S et al. DEAD-box RNA helicase subunits of the Drosha complex are required for processing of rRNA and a subset of microRNAs. Nat Cell Biol 2007; 9:604-611.
23. Wu H, Xu H, Miraglia LJ et al. Human RNase III is a 160-kDa protein involved in preribosomal RNA processing. J Biol Chem 2000; 275:36957-36965.
24. Jalal C, Uhlmann-Schiffler H, Stahl H. Redundant role of DEAD box proteins p68 (Ddx5) and p72/p82 (Ddx17) in ribosome biogenesis and cell proliferation. Nucleic Acids Res 2007; 35:3590-3601.
25. Kim VN, Han J, Siomi MC. Biogenesis of small RNAs in animals. Nat Rev Mol Cell Biol 2009; 10:126-139.
26. Lee Y, Kim M, Han J et al. microRNA genes are transcribed by RNA polymerase II. EMBO J 2004; 23:4051-4060.
27. Bracht J, Hunter S, Eachus R et al. Trans-splicing and polyadenylation of let-7 microRNA primary transcripts. RNA 2004; 10:1586-1594.
28. Cai X, Hagedorn CH, Cullen BR. Human microRNAs are processed from capped, polyadenylated transcripts that can also function as mRNAs. RNA 2004; 10:1957-1966.
29. Rodriguez A, Griffiths-Jones S, Ashurst JL et al. Identification of mammalian microRNA host genes and transcription units. Genome Res 2004; 14:1902-1910.
30. Kim YK, Kim VN. Processing of intronic microRNAs. EMBO J 2007; 26:775-783.

31. Bohnsack MT, Czaplinski K, Gorlich D. Exportin 5 is a RanGTP-dependent dsRNA-binding protein that mediates nuclear export of pre-miRNAs. RNA 2004; 10:185-191.
32. Yi R, Qin Y, Macara IG et al. Exportin-5 mediates the nuclear export of pre-microRNAs and short hairpin RNAs. Genes Dev 2003; 17:3011-3016.
33. Hutvagner G, McLachlan J, Pasquinelli AE et al. A cellular function for the RNA-interference enzyme Dicer in the maturation of the let-7 small temporal RNA. Science 2001; 293:834-838.
34. Grishok A, Pasquinelli AE, Conte D et al. Genes and mechanisms related to RNA interference regulate expression of the small temporal RNAs that control C. elegans developmental timing. Cell 2001; 106:23-34.
35. Ketting RF, Fischer SE, Bernstein E et al. Dicer functions in RNA interference and in synthesis of small RNA involved in developmental timing in C. elegans. Genes Dev 2001; 15:2654-2659.
36. Eulalio A, Tritschler F, Izaurralde E. The GW182 protein family in animal cells: new insights into domains required for miRNA-mediated gene silencing. RNA 2009; 15:1433-1442.
37. Carthew RW, Sontheimer EJ. Origins and Mechanisms of miRNAs and siRNAs. Cell 2009; 136:642-655.
38. Bousquet-Antonelli C, Presutti C, Tollervey D. Identification of a regulated pathway for nuclear pre-mRNA turnover. Cell 2000; 102:765-775.
39. Luke B, Panza A, Redon S et al. The Rat1p 5' to 3' exonuclease degrades telomeric repeat-containing RNA and promotes telomere elongation in Saccharomyces cerevisiae. Mol Cell 2008; 32:465-477.
40. Petfalski E, Dandekar T, Henry Y et al. Processing of the precursors to small nucleolar RNAs and rRNAs requires common components. Mol Cell Biol 1998; 18:1181-1189.
41. Henry Y, Wood H, Morrissey JP et al. The 5' end of yeast 5.8S rRNA is generated by exonucleases from an upstream cleavage site. EMBO J 1994; 13:2452-2463.
42. Geerlings TH, Vos JC, Raué HA. The final step in the formation of 25S rRNA in Saccharomyces cerevisiae is performed by 5'-->3' exonucleases. RNA 2000; 6:1698-1703.
43. Kim M, Krogan NJ, Vasiljeva L et al. The yeast Rat1 exonuclease promotes transcription termination by RNA polymerase II. Nature 2004; 432:517-522.
44. West S, Gromak N, Proudfoot NJ. Human 5' --> 3' exonuclease Xrn2 promotes transcription termination at co-transcriptional cleavage sites. Nature 2004; 432:522-525.
45. Luo W, Johnson AW, Bentley DL. The role of Rat1 in coupling mRNA 3'-end processing to transcription termination: implications for a unified allosteric-torpedo model. Genes Dev 2006; 20:954-965.
46. Banerjee A, Sammarco MC, Ditch S et al. A novel tandem reporter quantifies RNA polymerase II termination in mammalian cells. PLoS One 2009; 4:e6193.
47. Morlando M, Ballarino M, Gromak N et al. Primary microRNA transcripts are processed co-transcriptionally. Nat Struct Mol Biol 2008; 15:902-990.
48. Ballarino M, Pagano F, Girardi E et al. Coupled RNA processing and transcription of intergenic primary microRNAs. Mol Cell Biol 2009.
49. Gy I, Gasciolli V, Lauressergues D et al. Arabidopsis FIERY1, XRN2, and XRN3 are endogenous RNA silencing suppressors. Plant Cell 2007; 19:3451-3461.
50. Johnson AW. Rat1p and Xrn1p are functionally interchangeable exoribonucleases that are restricted to and required in the nucleus and cytoplasm, respectively. Mol Cell Biol 1997; 17:6122-6130.
51. Villa T, Ceradini F, Presutti C et al. Processing of the intron-encoded U18 small nucleolar RNA in the yeast Saccharomyces cerevisiae relies on both exo- and endonucleolytic activities. Mol Cell Biol 1998; 18:3376-3383.
52. Poole TL, Stevens A. Comparison of features of the RNase activity of 5'-exonuclease-1 and 5'-exonuclease-2 of Saccharomyces cerevisiae. Nucleic Acids Symp Ser 1995; 79-81.
53. Kenna M, Stevens A, McCammon M et al. An essential yeast gene with homology to the exonuclease-encoding XRN1/KEM1 gene also encodes a protein with exoribonuclease activity. Mol Cell Biol 1993; 13:341-350.
54. Stevens A, Poole TL. 5'-exonuclease-2 of Saccharomyces cerevisiae. Purification and features of ribonuclease activity with comparison to 5'-exonuclease-1. J Biol Chem 1995; 270:16063-16069.
55. Wang Y, Sheng G, Juranek S et al. Structure of the guide-strand-containing argonaute silencing complex. Nature 2008; 456:209-213.
56. Xiang S, Cooper-Morgan A, Jiao X et al. Structure and function of the 5'-->3' exoribonuclease Rat1 and its activating partner Rai1. Nature 2009; 458:784-788.
57. Chen Y, Pane A, Schüpbach T. Cutoff and aubergine mutations result in retrotransposon upregulation and checkpoint activation in Drosophila. Curr Biol 2007; 17:637-642.
58. Lu Y, Liu P, James M et al. Genetic variants cis-regulating Xrn2 expression contribute to the risk of spontaneous lung tumor. Oncogene 2009.
59. Büssing I, Slack FJ, Großhans H. let-7 microRNAs in development, stem cells and cancer. Trends Mol Med 2008; 14:400-409.

60. Zhou JY, Ma WL, Liang S et al. Analysis of microRNA expression profiles during the cell cycle in synchronized HeLa cells. BMB Rep 2009; 42:593-598.
61. Sethi P, Lukiw WJ. Micro-RNA abundance and stability in human brain: specific alterations in Alzheimer's disease temporal lobe neocortex. Neurosci Lett 2009; 459:100-104.
62. Schratt GM, Tuebing F, Nigh EA et al. A brain-specific microRNA regulates dendritic spine development. Nature 2006; 439:283-289.
63. Rajasethupathy P, Fiumara F, Sheridan R et al. Characterization of small RNAs in aplysia reveals a role for miR-124 in constraining synaptic plasticity through CREB. Neuron 2009; 63:803-817.
64. Iliopoulos D, Hirsch HA, Struhl K. An epigenetic switch involving NF-kappaB, Lin28, Let-7 microRNA, and IL6 links inflammation to cell transformation. Cell 2009; 139:693-706.
65. Newman MA, Thomson JM, Hammond SM. Lin-28 interaction with the Let-7 precursor loop mediates regulated microRNA processing. RNA 2008; 14:1539-1549.
66. Viswanathan SR, Daley GQ, Gregory RI. Selective blockade of microRNA processing by Lin28. Science 2008; 320:97-100.
67. Rybak A, Fuchs H, Smirnova L et al. A feedback loop comprising lin-28 and let-7 controls pre-let-7 maturation during neural stem-cell commitment. Nat Cell Biol 2008; 10:987-993.
68. Lehrbach NJ, Armisen J, Lightfoot HL et al. LIN-28 and the poly(U) polymerase PUP-2 regulate let-7 microRNA processing in Caenorhabditis elegans. Nat Struct Mol Biol 2009; 16:1016-1020.
69. Heo I, Joo C, Cho J et al. Lin28 mediates the terminal uridylation of let-7 precursor microRNA. Mol Cell 2008; 32:276-284.
70. Heo I, Joo C, Kim YK et al. TUT4 in concert with Lin28 suppresses microRNA biogenesis through pre-microRNA uridylation. Cell 2009; 138:696-708.
71. Hagan JP, Piskounova E, Gregory RI. Lin28 recruits the TUTase Zcchc11 to inhibit let-7 maturation in mouse embryonic stem cells. Nat Struct Mol Biol 2009; 16:1021-1025.
72. Katoh T, Sakaguchi Y, Miyauchi K et al. Selective stabilization of mammalian microRNAs by 3' adenylation mediated by the cytoplasmic poly(A) polymerase GLD-2. Genes Dev 2009; 23:433-438.
73. Sood P, Krek A, Zavolan M et al. Cell-type-specific signatures of microRNAs on target mRNA expression. Proc Natl Acad Sci USA 2006; 103:2746-2751.
74. Khan AA, Betel D, Miller ML et al. Transfection of small RNAs globally perturbs gene regulation by endogenous microRNAs. Nat Biotechnol 2009; 27:549-555.
75. Diederichs S, Haber DA. Dual role for argonautes in microRNA processing and post-transcriptional regulation of microRNA expression. Cell 2007; 131:1097-1108.
76. O'Carroll D, Mecklenbrauker I, Das PP et al. A Slicer-independent role for Argonaute 2 in hematopoiesis and the microRNA pathway. Genes Dev 2007; 21:1999-2004.
77. Kedde M, Strasser MJ, Boldajipour B et al. RNA-binding protein dnd1 inhibits microRNA access to target mRNA. Cell 2007; 131:1273-1286.
78. Huang J, Liang Z, Yang B et al. Derepression of microRNA-mediated protein translation inhibition by apolipoprotein B mRNA-editing enzyme catalytic polypeptide-like 3G (APOBEC3G) and its family members. J Biol Chem 2007; 282:33632-33640.
79. Ibrahim F, Rymarquis LA, Kim E et al. Uridylation of mature miRNAs and siRNAs by the MUT68 nucleotidyltransferase promotes their degradation in Chlamydomonas. Proc Natl Acad Sci USA 2010; 107:3906-3911.

INDEX

A

Adenosine deaminase 76
Adenosine deaminase acting on RNA
    (ADAR) 57, 76-83
Adenylation 9, 78, 132, 135, 150
Ago 6, 7, 24, 98, 100, 116, 117, 125,
    128-136, 140, 141, 143, 144, 147,
    151
Ago1 6, 7, 24, 93, 96-99
Ago2 7, 24, 99, 101
AGO4 129, 130
AIN-1 6, 96, 143
AIN-2 6, 143
APOBEC3G 118
Arabidopsis 6, 23, 124-126, 128-131,
    133-135, 141, 144, 152
Arabidopsis thaliana 124, 125, 129, 141
Argonaute 1-8, 10, 24, 96, 98-100, 116,
    125, 127-135, 140, 141, 144, 147
Asymmetric division 95
A-to-I editing 78-81
AU-rich element (ARE) 106-114, 116,
    117

B

Bardet-Biedl syndrome 88, 90
B-Box 85-88, 90-92
Berp 92

Bone morphogenetic protein (BMP) 15,
    16, 20, 21, 25, 111
Brain tumor (Brat) 85, 88, 89, 93-99,
    101, 102

C

C3PO 128
Caenorhabditis elegans 1, 2, 4-6, 67-73,
    83, 85-87, 91, 92, 96, 97, 99-101,
    118, 124, 126, 129-131, 133-135,
    141, 143-145, 148, 151, 152
Cancer 15, 19, 21, 22, 37, 39, 41, 48,
    49, 57-59, 64, 73, 74, 110-112, 148,
    149
Carcinoma 40, 58
Cell differentiation 40, 64, 67, 72, 92,
    100, 102
Cell fate 68, 70, 72, 93, 95, 97, 101
Cell proliferation 39, 40, 43, 46, 93, 98,
    107
Central nervous system (CNS) 46,
    77-79, 91-95, 98
CGH-1 96
Chlamydomonas 5, 124, 126, 131-134,
    136, 152
Chlamydomonas reinhardtii 124, 126,
    128, 129, 131, 132, 152
Chromatin 24, 46, 48, 125
Co-Smad 16, 17, 21, 25, 161

D

Dappled 88, 89, 91-93, 99, 101
DDX5 20, 37-39, 57, 141
DDX17 20, 37-39, 57, 141
DEAD-box 20, 29, 37, 44, 45, 57, 96, 126
DEAD-box RNA helicase 20, 37, 44, 57
Deadenylation 6, 112, 143
Development 4, 15, 18, 29, 40, 46, 56, 57, 59, 64, 67-70, 72-74, 85, 87, 89, 91-93, 95, 97, 99, 101, 102, 125, 141, 147, 148, 152
Developmental timing 17, 68, 69, 72, 86, 89, 91, 96
DGCR8 6, 19-22, 24, 28, 30, 39, 43, 46, 56-64, 80, 81, 83
Dicer 2-7, 10, 17, 18, 24, 28, 29, 32, 36-39, 41, 43, 49, 56, 57, 59-61, 63, 67, 68, 70-72, 80, 81, 83, 99, 100, 125-129, 132, 143, 144, 146, 147, 151, 152
DiGeorge syndrome 19, 57, 59
Disease 15, 19, 56-59, 64, 87, 102, 110, 112
Dnd1 118, 162
DRB1 130
Drosha 6, 10, 15, 17-24, 28-33, 36-39, 41, 43-46, 49-52, 56-64, 70, 71, 80-83, 126, 141, 143
Drosophila 2, 4-6, 57, 59, 61, 77, 85, 87, 91-96, 102, 107, 113, 125, 129, 149
Drosophila melanogaster 85, 86, 92, 97, 99, 133, 134
dsRNA 2, 3, 5, 77, 78, 82, 83, 125, 126, 128-130
dsRNA-binding protein 83, 129, 130

E

E2 22, 23, 46, 50, 87-90, 93, 99
E3-ligase 85, 87-90, 93, 97, 98, 99, 101
ELAV 106, 107, 113, 117, 118
Embryonic development 40, 91, 93
Embryonic stem cell (ES Cell) 40, 46, 60, 61, 64, 73, 99, 101

Endocytosis 93
ERI-1 130
ERα 22, 23, 44, 46-49, 52
ERβ 46, 47
Estradiol 22, 23, 37, 46
Estrogen 20, 22, 29, 37-39, 43, 44, 46-52, 58, 126, 150
Estrogen receptor 20, 22, 37-39, 43, 44, 46-48, 58, 126, 150
Exoribonuclease 124, 129-132, 134-136, 140, 143, 148, 157
Exosome 124, 131, 132, 144
Exportin-5 6, 17, 18, 28, 37, 38, 127, 130

F

Feedback loop 48, 56, 61, 100, 101, 111
5'→3' Exonuclease 129, 141, 143
5'UTR 59-62, 83, 111

G

Ganglion mother cell (GMC) 91, 94, 95
GLD2 150
Gonad 91, 95, 107
Growth factor 15, 29, 48, 92, 107, 108, 110
Guide strand 4, 7, 17, 37, 126, 143, 151
GW182 6, 8, 96, 100, 116, 143

H

Half-life 38, 111, 142, 149, 150
HASTY (HST) 6, 127, 130
Helicase 2, 20, 22, 23, 29, 37, 43-46, 48, 52, 57, 96, 126, 141
HEN1 127-130, 133, 134
hnRNP A1 28-33, 38, 40, 57, 82, 126
hnRNP protein 28-33, 38, 40, 82, 126
HuB 107, 110, 118
HuC 107, 118
HuD 107, 110, 111, 118
HuR 106-118
HYL1 6, 130

I

Ig-Filamin repeat   89
Inosine   76-78, 80, 81
Intron   6, 17, 36, 57, 77, 125, 126, 141,
    144
iPS cells   64, 72, 73

K

KH-type splicing regulatory protein
    (KHSRP)   37, 38, 68

L

Let-7   4, 5, 19, 39-41, 67-73, 78, 86,
    91, 92, 94, 96-102, 116-118, 126,
    145-152
Limb girdle muscular dystrophy (LGMD)
    88, 90, 91
Lin-4   4-6, 68-70, 86, 96, 146, 148
Lin-28   32, 39-41, 67-73, 78, 99-102,
    126
Lin-41   68, 85-87, 89, 91, 93, 95, 96,
    98-101
LooptomiR   32

M

Meiosis   95
Mei-P26   85, 89, 93-97, 99
Methyltransferase   48, 110, 128-130
Microprocessor   6, 19, 22, 23, 28, 29, 38,
    43, 56-64, 80, 81, 126
MicroRNase (miRNase)   140, 141, 143,
    152
miR*   7, 143
miR-18a   28-33, 38, 57
miR-21   19-23, 39
miR-22   80, 81
miR-122   78, 114-116, 118, 132, 150
miR-125   91, 99, 100, 112
miR-363   79
miRISC   93, 96-101, 143, 147, 151

miRNA   1, 3-10, 15, 17-25, 28-33,
    36-41, 43-46, 49-52, 56-64, 67-71,
    76-83, 85-87, 91, 95-102, 106, 111,
    112, 114-119, 124-136, 140-152
miRNA biogenesis   5, 6, 9, 15, 17, 18,
    20, 24, 25, 32, 33, 37, 39-41, 43, 56,
    57, 63, 64, 76, 78, 81-83, 96, 127,
    129, 141-143
miRNA expression atlas   79
miRNA maturation   19-22, 24, 37-41, 56
miRNA*   17, 18, 37, 126-130
miRNA processing   6, 15, 20-23, 28, 31,
    32, 36, 37, 40-46, 49-52, 57, 59, 60,
    63, 64, 67, 68, 126, 142, 143, 146,
    149
miRNA stability   149, 150
miRNA turnover   124, 140, 141, 143,
    144, 148-152
Mirtron   6, 57
mRNA stability   15, 48, 49, 61, 119
mRNA turnover   106, 111-113
Muscle   20, 78, 79, 89-92
MUT68   124, 129, 131-134, 136, 152
Myc   29, 40, 73, 94, 95, 97, 98, 100-102,
    112, 117, 118
Myosin   90, 92, 93

N

Neural activity-related ring finger
    (NARF)   93
Neuroblast   91, 93-96, 110
Neuron   91-94, 118, 149
Neuronal differentiation   97, 98, 101
NHL   85-102
NHL-2   85, 89, 96-101
NHL domain   85, 87-93, 95, 99
Nuclear receptor   46, 48
Nucleotidyltransferase   124, 129,
    131-133, 135, 152

O

Oncogene   57, 73, 85, 87, 117, 148
Oocyte   91, 94, 95

P

p68 20, 22, 23, 29, 37, 44-46, 48-52, 57,
    126, 141,
    see also DDX5
p72 20, 22, 23, 29, 37, 44-46, 48-50, 52,
    57, 126, 141,
    see also DDX17
PACT 6, 7
Pasha 6, 19, 57, 59, 61
Passenger strand 3, 4, 7, 17, 37, 125,
    126, 128, 143,
    see also miR*, miRNA*
P-body 9, 90, 96, 98, 99, 115-117
Peripheral nervous system (PNS) 91-93,
    95
Piwi 2, 4, 125, 128, 134
Pluripotency 73, 100, 101
Pluripotent stem cell 64, 72
Polyadenylation 111, 112
Poly (U) polymerase 71, 72, 126
Posttranscriptional regulation 31, 37, 48,
    59, 61, 62, 87, 91, 101, 112, 125
pre-let-7 39, 71, 78, 100, 145
pre-miR-151 81
pre-miRNA 17-19, 22, 23, 28, 37-39,
    43, 49, 56, 57, 59, 60, 67, 68, 70, 71,
    78-81, 83, 126, 127, 130, 141-147,
    149-152
pri-miR-142 80, 81
pri-miR-203 81
pri-miR-376a 79
pri-miRNA 6, 15, 17-20, 22, 23, 28,
    30-33, 36-40, 43-46, 49-52, 56, 57,
    59-63, 67, 78-83, 126, 127, 141-144,
    149
Proteasome 88, 90, 99
Pumilio 89, 93, 118
PUP-2 71, 72

R

Rat1p 141, 143, 144, 148
Rex1p 141
Ribonuclease (RNase) 2, 6, 15, 17, 19,
    28, 31, 37, 38, 43, 56, 57, 63, 96,
    125, 126, 128, 130, 132, 135, 141,
    143, 152

RING 85-89, 91, 92, 97-99, 101
RISC-loading complex (RLC) 7, 129
RNA-binding protein (RBP) 6, 7, 28,
    29, 32, 33, 37-40, 43, 44, 51, 52, 56,
    57, 61, 63, 67-69, 82, 89, 93, 100,
    106, 107, 112, 114, 116, 118, 119,
    126
RNA editing 76, 78, 79, 81-83
RNA-induced silencing complex (RISC)
    2, 4, 7, 8, 17-19, 24, 29, 37, 81, 125,
    127-136, 147
RNA interference (RNAi) 1-5, 7, 8,
    10, 24, 61, 64, 95, 115, 116, 124,
    130-134, 136, 144-148
RNA polymerase II (RNA pol II) 6, 17,
    18, 24, 36, 43, 49, 57, 108, 126, 127,
    141
RNase III 2, 6, 15, 17, 19, 28, 36, 37, 43,
    57, 63, 125, 126, 141, 143,
    see also Drosha, Dicer
RNA recognition motifs (RRM) 38, 44,
    107, 109, 112, 113
RNASEN 57,
    see also Drosha
RRP6 124, 129, 131-134, 152
R-Smad 16, 17, 20-23

S

Seam cell 68, 69, 91
Self-renewal 46, 73, 95, 96, 99, 101
Signaling 15
Smad 15-17, 20-25, 52, 57
Small interfering RNA (siRNA) 2-5, 7,
    61, 115, 124-126, 128-136, 151
Small RNA 1-5, 7, 8, 10, 60, 79, 99,
    124-126, 128-136, 140, 141
Small RNA degradation 126, 130, 132,
    135, 136
Small RNA degrading nuclease (SDN)
    124, 129, 131, 134, 140, 141, 152
Splicing 29, 34, 36-38, 44, 57, 143, 144
Stem cell 40, 64, 67, 68, 71-74, 85, 94,
    96, 98-102

T

Terminal loop (TL)   19, 28, 30-33, 36, 38-41, 70, 71
Tetrahymena thermophila   128
3'-to-5' exonuclease   78, 132, 141, 157
3'-to-5' exoribonuclease   129, 157
3'UTR   4, 8, 10, 18, 49, 106, 114-117, 145
TNRC6   6, 9, 143,
    see also GW182, AIN-1, AIN-2
Transforming growth factor β (TGFβ) 15-17, 20-22, 24, 25
Translational repression   96, 112, 116, 117, 119, 125, 143
Translin   128
Transposon   124, 125
TRBP   6, 7, 61, 63, 64, 80, 81, 83
Trim2   89, 92, 93, 95, 99, 101
Trim3   89, 92, 93, 101
Trim18   88
Trim32   85, 87-102
Trim71   88, 91
Tudor-SN   80, 81, 83
Tumor   19, 21, 22, 25, 29, 37, 41, 57-59, 63, 91, 93-95, 102, 109, 111, 148

Tumor suppressor   22, 37, 57, 93, 109, 111, 148
Tut4   71, 72
2'-O-methylation   78, 128, 136, 152

U

Ubiquitin   85, 87-90, 93, 97, 99-102
Uridylation   39, 67, 71, 72, 78, 100, 124, 126, 128, 131-133, 135, 152

V

Virus   79, 81, 112, 124, 125

W

Wech   88, 89, 91, 92, 101

X

XRN-1   144, 148, 152
XRN-2   124, 129-131, 134, 140, 141, 143-149, 151, 152